Principles of Ecosystem Management

Principles of Ecosystem Management

Jaxon Pine

STATES
ACADEMIC PRESS
www.statesacademicpress.com

States Academic Press,
109 South 5th Street,
Brooklyn, NY 11249, USA

Visit us on the World Wide Web at:
www.statesacademicpress.com

ISBN: 978-1-63989-436-9 (Hardback)

Cataloging-in-Publication Data

Principles of ecosystem management / Jaxon Pine.
p. cm.
Includes bibliographical references and index.
ISBN 978-1-63989-436-9
1. Ecosystem management. 2. Applied ecology. 3. Biodiversity conservation.
4. Environmental management. 5. Nature conservation. I. Pine, Jaxon.

QH75 .P75 2022
333.720 973--dc23

Table of Contents

Preface

It is with great pleasure that I present this book. It has been carefully written after numerous discussions with my peers and other practitioners of the field. I would like to take this opportunity to thank my family and friends who have been extremely supporting at every step in my life.

The conservation of ecological services and efficient restoration of natural resources is termed as ecosystem management. It aims to meet the socioeconomic, political and ecological needs of the current and future generation. Natural resource management is the sustainable utilization of natural resources such as land, water, soil plants, animals, etc. Adaptive resource management is a learning process which improves management outcomes on a long-run. Its ultimate goal is the management of ecosystem to maintain the ecological integrity. Other approaches to ecosystem management include strategic management, landscape level conservation, and command and control management. This book is a valuable compilation of topics, ranging from the basic to the most complex theories and principles in the field of ecosystem management. It elucidates new techniques and their applications in a multidisciplinary approach. In this book, constant effort has been made to make the understanding of the difficult concepts of ecosystem management as easy and informative as possible, for the readers.

The chapters below are organized to facilitate a comprehensive understanding of the subject:

Chapter - What is Ecology?

The branch of biology that is concerned with the study of interactions between the organisms and their biophysical environment is called ecology. Some of the branches of ecology are microbial ecology, forest ecology, terrestrial ecology, behavioral ecology, population ecology and applied ecology. The topics elaborated in this chapter will help in gaining a better perspective of ecology and its branches.

Chapter - Ecosystems and Ecoregions

The community of living organisms along with the non-living components of their environment is called an ecosystem. The major types of ecosystems are forest ecosystem, desert ecosystem, aquatic ecosystem, etc. Some of the ecoregions are phytochorion, forest steppe, alpine-steppe, etc. This chapter has been carefully written to provide an easy understanding of these types of ecosystem and ecoregions.

Chapter - Understanding Ecosystem Management

The process of management that is aimed at conserving ecological services and restoring natural resources is termed as ecosystem management. Conservation biology, soil conservation, watershed management, etc. are some of the areas studied under its domain. All these aspects related to ecosystem management have been carefully analyzed in this chapter.

Chapter – Natural Resources and their Management

Natural resources are categorized into renewable and non-renewable resources. Renewable resources include sunlight, air, water, etc. whereas non-renewable resources include coal, petroleum and fossil fuels. Management of these resources for maintaining the quality of life is referred to as natural resource management. This chapter discusses natural resources and their management in detail.

Chapter – Techniques for Ecological Conservation

Some of the techniques that are used for ecological conservation are bush regeneration, conservation grazing, phytoremediation, prairie and mangrove restoration, gradsect, controlled burning, etc. This chapter closely examines these techniques of ecological conservation for an extensive understanding of the subject.

Jaxon Pine

1

What is Ecology?

The branch of biology that is concerned with the study of interactions between the organisms and their biophysical environment is called ecology. Some of the branches of ecology are microbial ecology, forest ecology, terrestrial ecology, behavioral ecology, population ecology and applied ecology. The topics elaborated in this chapter will help in gaining a better perspective of ecology and its branches.

Ecology or ecological science, is the scientific study of the distribution and abundance of living organisms and how these properties are affected by interactions between the organisms and their environment. The environment of an organism includes both the physical properties, which can be described as the sum of local abiotic factors like climate and geology, as well as the other organisms that share its habitat.

Ecology may be more simply defined as the relationship between living organisms and their abiotic and biotic environment or as "the study of the structure and function of nature". In this later case, structure includes the distribution patterns and abundance of organisms, and function includes the interactions of populations, including competition, predation, symbiosis, and nutrient and energy cycles.

Ecology is essentially the study of the Earth's organisms and their relationships with their abiotic and biotic environments.

The interactions between living organisms and their abiotic and biotic environments, the focus of ecology, generally convey an overall sense of unity and harmony in nature. On the other hand, the history of the science itself has often revealed conflicts, schisms, and opposing camps, as ecologists took different approaches and often failed to meet on common ground.

Ecology is usually considered a branch of biology, the general science that studies living and once-living organisms. Organisms can be studied at many different levels, from proteins and nucleic acids (in biochemistry and molecular biology), to cells (in cellular biology), to multicellular systems (in physiology and anatomy, to individuals (in botany, zoology, and other similar disciplines), and finally at the level of populations, communities, and ecosystems, and to the biosphere as a whole. These latter strata, from populations to the biosphere, are the primary subjects of ecological inquiries.

The complexity of the subject of ecological studies can be see in this river in South America, which is part of the world's largest wetland, the Pantanal.

Ecology is a multi-disciplinary science. Because of its focus on the higher levels of the organization of life on earth and on the interrelations between organisms and their environment, ecology draws heavily on many other branches of science, especially geology and geography, meteorology, pedology, chemistry, and physics. Thus, ecology is said to be a holistic science, one that overarches older disciplines, such as biology, which in this view become sub-disciplines contributing to ecological knowledge.

Agriculture, fisheries, forestry, medicine, and urban development are among human activities that would fall within Krebs explanation of his definition of ecology: "where organisms are found, how many occur there, and why."

The term ecology is sometimes confused with the term environmentalism. Environmentalism is a social movement aimed at the goal of protecting natural resources or the environment, and which may involve political lobbying, activism, education, and so forth. Ecology is the science that studies living organisms and their interactions with the environment. As such, ecology involves scientific methodology and does not dictate what is "right" or "wrong." However, findings in ecology may be used to support or counter various goals, assertions, or actions of environmentalists.

Consider the ways an ecologist might approach studying the life of honeybees:

- The behavioral relationship between individuals of a species is behavioural ecology for example, the study of the queen bee, and how she relates to the worker bees and the drones.
- The organized activity of a species is community ecology; for example, the activity of bees assures the pollination of flowering plants. Bee hives additionally produce honey, which is consumed by still other species, such as bears.
- The relationship between the environment and a species is environmental ecology for example, the consequences of environmental change on bee activity. Bees may die out due to environmental changes. The environment simultaneously affects and is a consequence of this activity and is thus intertwined with the survival of the species.

Basic Concepts in Ecology

Ecology is a very broad-ranging and complex topic, and even its definition lacks consensus. Thus, there are numerous concepts that fit within this discipline, and diverse manners in which the content can be arranged and studied. Several of the basic concepts of ecology include ecological units, the ecosystem, energy flow, nutrient cycles, species interaction, productivity, and ecological challenges.

Ecological Units

For modern ecologists, ecology can be studied at several levels: population level (individuals of the same species), biocenosis level (or community of species), ecosystem level, biome level, and biosphere level.

The outer layer of the planet Earth can be divided into several compartments: the hydrosphere (or sphere of water), the lithosphere (or sphere of soils and rocks), and the atmosphere (or sphere of the air). The biosphere (or sphere of life), sometimes described as "the fourth envelope," is all living matter on the planet or that portion of the planet occupied by life. It reaches well into the other three spheres, although there are no permanent inhabitants of the atmosphere. Most life exists on or within a few meters of the Earth's surface. Relative to the volume of the Earth, the biosphere is only the very thin surface layer that extends from 11,000 meters below sea level to 15,000 meters above.

It is thought that life first developed in the hydrosphere, at shallow depths, in the photic zone (the area of water exposed to sufficient sunlight for photosynthesis). Multicellular organisms then appeared and colonized benthic zones. Terrestrial life developed later, after the ozone layer protecting living beings from UV rays formed. Diversification of terrestrial species is thought to be increased by the continents drifting apart, or alternately, colliding. Biodiversity is expressed at the ecological level (ecosystem), population level (intraspecific diversity), species level (specific diversity), and genetic level. Recently, technology has allowed the discovery of the deep ocean vent communities. This remarkable ecological system is not dependent on sunlight but bacteria, utilizing the chemistry of the hot volcanic vents, as the base of its food chain.

The biosphere contains great quantities of elements such as carbon, nitrogen, and oxygen. Other elements, such as phosphorus, calcium, and potassium, are also essential to life, yet are present in smaller amounts. At the ecosystem and biosphere levels, there is a continual recycling of all these elements, which alternate between their mineral and organic states.

A biome is a homogeneous ecological formation that exists over a vast region, such as tundra or steppes. The biosphere comprises all of the Earth's biomes the entirety of places where life is possible from the highest mountains to the depths of the oceans.

Biomes correspond rather well to subdivisions distributed along the latitudes, from the equator towards the poles, with differences based on the physical environment (for example, oceans or mountain ranges) and on the climate. Their variation is generally related to the distribution of species according to their ability to tolerate temperature and/or dryness. For example, one may find photosynthetic algae only in the photic part of the ocean (where light penetrates), while conifers are mostly found in mountains.

Though this is a simplification of a more complicated scheme, latitude and altitude approximate a good representation of the distribution of biodiversity within the biosphere. Very generally, biodiversity is greater near the equator (as in Brazil) and decreases as one approaches the poles.

The biosphere may also be divided into ecozones, which are biogeographical and ecological land classifications, such as Neartic, Neotropic, and Oceanic. Biozones are very well defined today and primarily follow the continental borders.

Ecological factors that can affect dynamic change in a population or species in a given ecology or environment are usually divided into two groups: biotic and abiotic.

Biotic factors relate to living organisms and their interactions. A biotic community is an assemblage of plant, animal, and other living organisms.

Abiotic factors are geological, geographical, hydrological, and climatological parameters. A biotope is an environmentally uniform region characterized by a particular set of abiotic ecological factors. Specific abiotic factors include:

- Water, which is at the same time an essential element to life and a milieu.
- Air, which provides oxygen, nitrogen, and carbon dioxide to living species and allows the dissemination of pollen and spores.

- Soil, at the same time a source of nutriment and physical support (soil pH, salinity, nitrogen, and phosphorus content, ability to retain water and density are all influential).
- Temperature, which should not exceed certain extremes, even if tolerance to heat is significant for some species.
- Light, which provides energy to the ecosystem through photosynthesis.
- Natural disasters can also be considered abiotic.

The Ecosystem Concept

Ecologists study ecosystems such as this section of the Salobra River in the Pantanal area of Brazil.

Some consider the ecosystem (abbreviation for "ecological system") to be the basic unit in ecology. An ecosystem is an ecological unit consisting of a biotic community together with its environment. Examples include a swamp, a meadow, and a river. It is generally considered smaller than a biome ("major life zone"), which is a large, geographic region of the earth's surface with distinctive plant and animal communities. A biome is often viewed as a grouping of many ecosystems sharing similar features, but is sometimes defined as an extensive ecosystem spread over a wide geographic area.

The first principle of ecology is that each living organism has an ongoing and continual relationship with every other element that makes up its environment. The ecosystem is composed of two entities, the entirety of life (the community, or biocoenosis) and the medium that life exists in (the biotope). Within the ecosystem, species are connected and dependent upon one another in the food chain, and exchange energy and matter between themselves and with their environment.

The concept of an ecosystem can apply to units of variable size, such as a pond, a field, or a piece of deadwood. A unit of smaller size is called a microecosystem. For example, an ecosystem can be a stone and all the life under it. A mesoecosystem could be a forest, and a macroecosystem a whole ecoregion, with its watershed.

Some of the main questions when studying an ecosystem include:

- How could the colonization of a barren area be carried out?
- What are the ecosystem's dynamics and changes?
- How does an ecosystem interact at local, regional, and global scale?
- Is the current state stable?
- What is the value of an ecosystem? How does the interaction of ecological systems provide benefit to humans, especially in the provision of healthy water?

Ecosystems are not isolated from each other, but are interrelated. For example, water may circulate between ecosystems by the means of a river or ocean current. Water itself, as a liquid medium, even defines ecosystems. Some species, such as salmon or freshwater eels move between marine systems and fresh-water systems. These relationships between the ecosystems lead to the concept of a biome.

Energy Flow

One focus of ecologists is to study the flow of energy, a major process linking the abiotic and biotic constituents of ecosystems.

While there is a slight input of geothermal energy, the bulk of the functioning of the ecosystem is based on the input of solar energy. Plants and photosynthetic microorganisms convert light into chemical energy by the process of photosynthesis, which creates glucose (a simple sugar) and releases free oxygen. Glucose thus becomes the secondary energy source that drives the ecosystem. Some of this glucose is used directly by other organisms for energy. Other sugar molecules can be converted to other molecules such as amino acids. Plants use some of this sugar, concentrated in nectar, to entice pollinators to aid them in reproduction.

Cellular respiration is the process by which organisms (like mammals) break the glucose back down into its constituents, water and carbon dioxide, thus regaining the stored energy the sun originally gave to the plants. The proportion of photosynthetic activity of plants and other photosynthesizers to the respiration of other organisms determines the specific composition of the Earth's atmosphere, particularly its oxygen level. Global air currents mix the atmosphere and maintain nearly the same balance of elements in areas of intense biological activity and areas of slight biological activity.

Nutrient Cycles

Ecologists also study the flow of nutrients in ecosystems. Whereas energy is not cycled, nutrients are cycled. Living organisms are composed mainly of carbon, oxygen, hydrogen, and nitrogen, and these four elements are cycled through the biotic communities and the geological world. These permanent recyclings of the elements are called biogeochemical cycles. Three fundamental biogeochemical cycles are the nitrogen cycle, the water cycle, and the carbon-oxygen cycle. Another key cycle is the phosphorus cycle.

Water is also exchanged between the hydrosphere, lithosphere, atmosphere, and biosphere. The oceans are large tanks that store water; they ensure thermal and climatic stability, as well as the transport of chemical elements thanks to large oceanic currents.

Species Interactions

Biocenose, or community, is a group of populations of plants, animals, and microorganisms. Each population is the result of procreations between individuals of same species and cohabitation in a given place and for a given time. When a population consists of an insufficient number of individuals, that population is threatened with extinction; the extinction of a species can approach when all biocenoses composed of individuals of the species are in decline. In small populations, consanguinity (inbreeding) can result in reduced genetic diversity that can further weaken the biocenose.

Mutual symbiosis between clownfish that dwell among the tentacles of tropical sea anemones, protects the anemone from anemone-eating fish, and in turn the stinging tentacles of the anemone protects the clownfish from its predators.

Biotic ecological factors influence biocenose viability; these factors are considered as either intraspecific or interspecific relations.

Intraspecific relations are those which are established between individuals of the same species, forming a population. They are relations of cooperation or competition, with division of the territory, and sometimes organization in hierarchical societies.

Interspecific relations interactions between different species are numerous, and are usually described according to their beneficial, detrimental, or neutral effect (for example, mutualism or competition). Symbiosis refers to an interaction between two organisms living together in more or less intimate association. A significant relation is predation (to eat or to be eaten), which leads to the essential concepts in ecology of food chains (for example, the grass is consumed by the herbivore, itself consumed by a carnivore, itself consumed by a carnivore of larger size). A high predator-to-prey ratio can have a negative influence on both the predator and prey biocenoses in that low availability of food and high death rate prior to sexual maturity can decrease (or prevent the increase of) populations of each, respectively. Other interspecific relations include parasitism, infectious disease, and competition for limiting resources, which can occur when two species share the same ecological niche.

In an ecosystem, the connections between species are generally related to food and their role in the food chain. There are three categories of organisms:

- Producers: Plants which are capable of photosynthesis.
- Consumers: Animals, which can be primary consumers (herbivorous), or secondary or tertiary consumers (carnivorous).
- Decomposers: Bacteria, mushrooms, which degrade organic matter of all categories, and restore minerals to the environment.

These relations form sequences in which each individual consumes the preceding one and is consumed by the one following, in what are called food chains or food networks.

The existing interactions between the various living beings go along with a permanent mixing of mineral and organic substances, absorbed by organisms for their growth, their maintenance, and their reproduction, to be finally rejected as waste. The interactions and biogeochemical cycles create a durable stability of the biosphere (at least when unchecked human influence and extreme weather or geological phenomena are left aside). This self-regulation, supported by negative feedback controls, supports the perenniality of the ecosystems. It is shown by the very stable concentrations of most elements of each compartment. This is referred to as homeostasis.

The ecosystem also tends to evolve to a state of ideal balance, reached after a succession of events, the climax (for example, a pond can become a peat bog).

Overall, the interactions of organisms convey a sense of unity and harmony. Plants, through photosynthesis, use carbon dioxide and provide oxygen, while animals use oxygen and give off carbon dioxide. On the level of the food web, plants capture the sun's energy and serve as food for herbivores, which serve as food for carnivores, and ultimately top carnivores. Decomposers (bacteria, fungi, etc.) break down organisms after they die into minerals that can be used by plants.

The harmony of species interactions with other species and the environment, including the biogeochemical cycles, have proposed a theory by some that the entire planet acts as if one, giant, functioning organism (the Gaia theory).

The observed harmony can be attributed to the concept of dual purpose: the view that every entity in the universe in its interactions simultaneously exhibits purposes for the whole and for the individual and that these purposes are interdependent. “Individual purpose” refers to the individual’s requirement to met basic needs of self-preservation, self-strengthening, multiplication, and development. The “whole purpose” is that by which the individual contributes to the preservation, strengthening, and development of the larger entity of which it is a part. Thus, the cell of a multicellular body provides a useful function for the body of which it is part. This “whole purpose,” which could be the secretion of an enzyme, harmonizes with the body’s requirement of self-preservation, development, self-strengthening, and reproduction. The body, on the other hand, supports the cell’s “individual purpose” by providing essential nutrients and carrying away wastes, assisting the cell’s self-preservation, self-strengthening, multiplication, and development. Likewise, each individual organism exhibits both an individual purpose and a purpose for the whole related to its place in the environment. The result is an extraordinary harmony evident in creation.

Ecosystem Productivity

The concepts dealing with the movement of energy through an ecosystem (via producers, consumers, and decomposers) lead to the idea of biomass (the total living matter in a given place), of primary productivity (the increase in the mass of plants during a given time), and of secondary productivity (the living matter produced by consumers and the decomposers in a given time).

A meandering river.

These two last ideas are key, since they make it possible to evaluate the load capacity the number of organisms which can be supported by a given ecosystem. In any food network, the energy contained in the level of the producers is not completely transferred to the consumers. Thus, from an energy point of view, it is more efficient for humans to be primary consumers (to get nourishment from grains and vegetables) than as secondary consumers (from herbivores such as beef and veal), and more still than as tertiary consumers (from eating carnivores).

The productivity of ecosystems is sometimes estimated by comparing three types of land-based ecosystems and the total of aquatic ecosystems:

- The forests (one-third of the Earth's land area) contain dense biomasses and are very productive. The total production of the world's forests corresponds to half of the primary production.
- Savannas, meadows, and marshes (one-third of the Earth's land area) contain less dense biomasses, but are productive. These ecosystems represent the major part of what humans depend on for food.
- Extreme ecosystems in the areas with more extreme climates deserts and semi-deserts, tundra, alpine meadows, and steppes (one-third of the Earth's land area) have very sparse biomasses and low productivity.
- Finally, the marine and fresh water ecosystems (three-fourths of Earth's surface) contain very sparse biomasses (apart from the coastal zones).

Humanity's actions over the last few centuries have reduced the amount of the Earth covered by forests (deforestation), and have increased agro-ecosystems (agriculture). In recent decades, an increase in the areas occupied by extreme ecosystems has occurred (desertification).

Ecological Challenges

Generally, an ecological crisis is what occurs when the environment of a species or a population evolves in a way unfavorable to that species survival.

It may be that environment quality degrades compared to the species needs, after a change in an abiotic ecological factor (for example, an increase of temperature, less significant rainfalls). It may be that the environment becomes unfavorable for the survival of a species (or a population) due to an increased pressure of predation (e.g., overfishing). It may be that the situation becomes unfavorable to the quality of life of the species (or the population) due to a rise in the number of individuals (overpopulation).

Although ecological crises are generally considered to be something that occurs in a short time span (days, weeks, or years), by definition, ecological crises can also be considered to occur over a very long time period, such as millions of years. They can also be of natural or anthropic origin. They may relate to one unique species or to many species.

People often use fires to clear land for agriculture or grazing livestock in the Pantanal.

Lastly, an ecological crisis may be local (an oil spill, a fire, or eutrophication of a lake), widespread (the movement of glaciers during an ice age), or global (a rise in the sea level).

According to its degree of endemism, a local crisis will have more or less significant consequences, from the death of many individuals to the total extinction of a species. Whatever its origin, disappearance of one or several species often will involve a rupture in the food chain, further impacting the survival of other species. Of course, what is an ecological crisis to one species, or one group of species, may be beneficial or neutral with respect to other species, at least short-term.

In the case of a global crisis, the consequences can be much more significant; some extinction events showed the disappearance of more than 90 percent of existing species at that time. However, it should be noted that the disappearance of certain species, such as the dinosaurs, by freeing an ecological niche, allowed the development and the diversification of the mammals. An ecological crisis may benefit other species, genera, families, orders, or phyla of organisms.

Sometimes, an ecological crisis can be a specific and reversible phenomenon at the ecosystem scale. But more generally, the crisis' impact will last. Indeed, it rather is a connected series of events that occur until a final point. From this stage, no return to the previous stable state is possible, and a new stable state will be set up gradually.

Lastly, if an ecological crisis can cause extinction, it can also more simply reduce the quality of life of the remaining individuals. Thus, even if the diversity of the human population is sometimes considered threatened (see in particular indigenous people), few people envision human disappearance at short span. However, epidemic diseases, famines, impact on health of reduction of air quality, food crises, reduction of living space, accumulation of toxic or non-degradable wastes, threats on key species (great apes, pandas, whales) are also factors influencing the well-being of people.

During the past decades, this increasing responsibility of humanity in some ecological crises has been clearly observed. Due to the increases in technology and a rapidly increasing population, humans have more influence on their own environment than any other ecosystem engineer.

Some usually quoted examples as ecological crises are:

- Permian-Triassic extinction event—250 million of years ago.
- Cretaceous-Tertiary extinction event—65 million years ago.
- Ozone layer hole issue.
- Deforestation and desertification, with the disappearance of many species.
- The nuclear meltdown at Chernobyl in 1986 that caused the death of many people and animals from cancer, and caused mutations in a large number of animals and people. The area around the plant is now abandoned because of the large amount of radiation generated by the meltdown.

2 Ecosystems and Ecoregions

The community of living organisms along with the non-living components of their environment is called an ecosystem. The major types of ecosystems are forest ecosystem, desert ecosystem, aquatic ecosystem, etc. Some of the ecoregions are phytochorion, forest steppe, alpine-steppe, etc. This chapter has been carefully written to provide an easy understanding of these types of ecosystem and ecoregions.

Ecosystem

In general terms, an ecosystem can be defined as an ecological unit consisting of a biotic community (an assemblage of plant, animal, and other living organisms) together with its abiotic environment (such as soil, precipitation, sunlight, temperature, slope of the land, etc.). The word ecosystem is an abbreviation of the term, "ecological system." A river, a swamp, a meadow, and a cave are all examples of ecosystems. Some consider the ecosystem to be the basic unit in ecology.

Humans are part of the environment and thus impact, and are impacted by, ecosystems. Forests provide wood for homes and an environment for recreation; wetlands purify our water; rivers provide fish and hydroelectric energy. Fifty percent of all jobs worldwide are tied to agriculture, forestry, and fishing. Human impacts often have caused dramatic changes to diverse ecosystems. Urbanization and industrial, agriculture, recreational, and forestry activities have impacted such things as biodiversity and numbers of organisms, modified biogeochemical cycles, and increased pollution.

The twentieth century exhibited humanity's ingenuity in many ways, including a history of intervening in major river and wetland systems by creating dams for hydroelectric plants or navigation, or by diverting water to open up wetlands for development. All large rivers in the temperate zone have been altered for human use, as have most of the world's large river floodplain ecosystems. It has been said that historically, if a forest, wetland, or river was not producing jobs and wealth, it was cut, drained, mined, or dammed. Clearly, the study of ecosystems and human impacts is important for creating a sustainable environment for future generations.

Size, Scale and Boundaries

The size and scale of an ecosystem can vary widely. They may be very large, such as a tropical rain forest, the Everglades, or the Pantanal, or very small, such as a test tube of phytoplankton or an aquarium tank with plants and fish. Some even define a biome as an extensive ecosystem, although generally an ecosystem is viewed as having a more defined abiotic environment than a biome, and a biome as a group of ecosystems sharing broad environmental characteristics.

The boundary of an ecosystem is not always easy to delineate. Different ecosystems are often separated by geographical barriers, like deserts, mountains, or oceans, or are isolated otherwise, like lakes or rivers. As these borders are never rigid, ecosystems tend to blend into each other. For example, the boundary of a river may seem clear, yet caimans crawl from the river to bask in the sun, herons get food from the river but nest in trees, and tapirs may swim in the water and yet live on the land. To some extent, the whole earth can be seen as a single ecosystem, or a lake can be divided into several ecosystems, depending on the scale used.

Energy Flow

A major process linking the abiotic and biotic constituents of ecosystems is the flow of energy.

The main source of energy in almost all natural ecosystems is radiant energy from the sun. Primary producers or autotrophic organisms, such as plants, algae, and photosynthetic bacteria, take radiant energy and fix it into organic molecules by photosynthesis, such a creating glucose from carbon dioxide. Only a small portion of radiant energy actually is converted into biochemical form via photosynthesis. Studies suggest that ecosystems generally fix 3 percent or less of sunlight, and that for most ecosystems this figure is probably less than 1 percent. There are also other autotrophic organisms, such as chemosynthetic bacteria living around deep-sea vents that can manufacture their own food from chemical energy.

Energy then flows through the system when organisms eat each other. The trophic level, or feeding level, is a way of delineating the position of an organism in the food chain, that is, the relationship between what the organism eats and what it is eaten by. Autotrophs are at the base of food chains. Heterotrophs utilize the energy fixed in organic molecules by autotrophs. Herbivores, or primary consumers, are heterotrophs that eat autotrophs, such as antelopes that feed on grass or zooplankton that feed on phytoplankton in the ocean or in lakes. Carnivores are heterotrophs that eat herbivores or other carnivores, and include coyotes, tigers, owls, and preying mantises. Carnivores can be secondary consumers (those that eat an herbivore), or tertiary consumers (those that eat a carnivore that has eaten a herbivore), and so on. Omnivores are heterotrophs that consume either autotrophs (primary producers) or consumers (herbivores and carnivores), and include bears and humans. Scavengers, such as crows, are heterotrophs that feed on recently dead organisms. Decomposers are heterotrophs that obtain energy by breaking down dead organisms into their inorganic form, such as bracket fungi that break down dead tissues and wastes into carbon, nitrogen, and other inorganic compounds and elements. Autotrophs can then utilize these materials and use them in manufacturing food.

Energy flows through an ecosystem in the form of carbon-carbon bonds. As carbon-carbon bonds are broken, energy is released, which then can be used by the organism or dissipated as heat. Although energy flows through an ecosystem, only a portion of the energy available to an organism is actually stored by the organism, and thus the total energy in one trophic level never flows to the next level. That is, lower trophic levels always contain more total energy than higher trophic levels. Energy does not recycle, but ultimately all energy that is brought into an ecosystem is lost as heat.

A food chain identifies the sequence in which organisms obtain energy and feed in an ecosystem, such as from grass to insect to mouse to owl to scavenging vulture to decomposing bacteria. A food web shows a more complex relationship of feeding and energy flow among species in an ecosystem.

Nutrient Cycles

A second major process linking the biotic and abiotic constituents of an ecosystem is the flow of nutrients. Unlike energy, which is not cycled, inorganic nutrients are cycled in ecosystems. A biogeochemical cycle is the process by which inorganic materials, such as water, oxygen, carbon, calcium, and nitrogen, move through both the biotic communities (organisms) and the geological world (atmosphere, oceans, soil, and even rocks).

Four of the most important inorganic nutrient cycles are the water cycle, the nitrogen cycle, the phosphorus cycle, and the oxygen-carbon cycle.

For example, in the nitrogen cycle, although about 78 percent of the atmosphere is nitrogen gas, most living organisms cannot use atmospheric nitrogen. There is a process that converts atmospheric nitrogen into compounds that plants can use, such as nitrites and nitrates. The nitrogen cycle includes four major processes. "Nitrogen fixation" is the process whereby bacteria convert nitrogen gas into ammonia compounds. The "nitrification process" involves chemosynthetic bacteria oxidizing ammonia compounds to produce nitrites and nitrates (which can also enter the soil from other sources, such as a bolt of lightning or erosion of certain rocks). Plants can utilize nitrites and nitrates to form amino acids. In the "ammonification process," bacteria break down nitrogen-contain amino acids from dead organisms or their wastes and form ammonia compounds (which, again, can cycle to plants via the nitrification process). In "denitrification," anaerobic bacteria break down nitrates, releasing nitrogen gas back into the atmosphere

Threats to Ecosystem

Anything that attempts to alter the balance of the ecosystem potentially threatens the health and existence of that ecosystem. Some of these threats are not overly worrying as they may be naturally resolved provided the natural conditions are restored. Other factors can destroy ecosystems and render all or some of its life forms extinct. Here are a few:

Habitat Destruction

Economic activities such as logging, mining, farming and construction often involve clearing out places with natural vegetative cover. Very often, tampering with one factor of the ecosystem can have a ripple effect on it and affect many more or all other factors of that ecosystem. For example, clearing a piece of forest for timber can expose the upper layers of the soil to the sun's heat, causing erosion and drying. It can cause a lot of animals and insects that depended on the shade and moisture from the tree to die or migrate to other places.

Pollution

Water, land and air pollution all together play a crucial role in the health of ecosystems. Pollution may be natural or human-caused, but regardless they potentially release destructive agents or chemicals (pollutants) into the environments of living things. "In a lake, for example, it can create havoc on the ecological balance by stimulating plant growth and causing the death of fish due to suffocation resulting from lack of oxygen. The oxygen cycle will stop, and the polluted water will also affect the animals dependent on the lake water".

Eutrophication

This is the enrichment of water bodies with plant biomass as a result of the continuous inflow of nutrients particularly nitrogen and phosphorus. Eutrophication of water fuels excessive plant and algae growth and also hurts water life, often resulting in the loss of flora and fauna diversity. "The known consequences of cultural eutrophication include blooms of blue-green algae (i.e., cyanobacteria), tainted drinking water supplies, degradation of recreational opportunities, and hypoxia. The estimated cost of damage mediated by eutrophication in the U.S. alone is approximately $2.2 billion annually.

Invasive Species

Any foreign species (biological) that finds its way into an ecosystem, either by natural or human introduction can have an effect on the ecosystem. If this alien has the ability to prey on vulnerable and native members of that ecosystem, they will be wiped out, sooner or later. One devastating impact of introducing alien Nile Perch and Nile Tilapia into Lake Victoria in the 1970s was the extinction of almost half of the 350+ endemic species of fish in the cichlid family.

Overharvesting

Fish species, game and special plants all do fall victim from time to time as a result of over-harvesting or humans over-dependence on them. Overharvesting leads to the reduction in populations, community structures and distributions, with an overall reduction in recruitment. Lots of fish species are known to have reached their maximum exploitation level, and others will soon be. "For example, Oreochromis karongae is one of the most valuable food fishes in Malawi, but populations collapsed in the 1990s due to overfishing, and it is now assessed as Endangered".

UV Radiation

The sun's rays play an important role in living things. UV rays come in three main wavelengths: UVA, UVB and UVC, and they have different properties. UVA has long wavelengths and reaches the earth's surface all the time. It helps generate vitamin D for living things. UVB and UVC are more destructive and can cause DNA and cell damage to plans and animals. Ozone depletion is one way that exposes living things to UVB and UVC and the harm caused can wipe lots of species, and affect ecosystems members including humans.

Forest Ecosystem

Forest is a complex ecological system in which trees are the dominant life-form.

Tree-dominated forests can occur wherever the temperatures rise above 10 °C (50 °F) in the warmest months and the annual precipitation is more than 200 mm (8 inches). They can develop under a variety of conditions within these climatic limits, and the kind of soil, plant, and animal life differs according to the extremes of environmental influences. In cool, high-latitude subpolar regions, forests are dominated by hardy conifers like pines, spruces, and larches. These taiga (boreal) forests have prolonged winters and between 250 and 500 mm (10 and 20 inches) of rainfall annually. In more temperate high-latitude climates, mixed forests of both conifers and broad-leaved deciduous trees predominate. Broad-leaved deciduous forests develop in middle-latitude climates, where there is an average temperature above 10° C (50° F) for at least six months every year and annual precipitation is above 400 mm (16 inches). A growing period of 100 to 200 days allows deciduous forests to be dominated by oaks, elms, birches, maples, beeches, and aspens. In the humid climates of the equatorial belt, tropical rainforests develop. There heavy rainfall supports evergreens that have broad leaves instead of needle leaves, as in cooler forests. In the lower latitudes of the Southern Hemisphere, the temperate deciduous forest reappears.

Boreal forest, Alaska, U.S., dominated by spruce trees (Picea).

Mixed evergreen and hardwood forest on the slopes of the Adirondack Mountains near Keene Valley, New York.

Forest types are distinguished from each other according to species composition (which develops in part according to the age of the forest), the density of tree cover, type of soils found there, and the geologic history of the forest region.

Soil conditions are distinguished according to depth, fertility, and the presence of perennial roots. Soil depth is important because it determines the extent to which roots can penetrate into the earth and, therefore, the amount of water and nutrients available to the trees. The soil of taiga forests is sandy and quickly drained. Deciduous forests have brown soil, richer than sand in nutrients, and less porous. Rainforests and savanna woodlands have a soil layer rich in iron or aluminum, which give the soils either a reddish or yellowish cast. The amount of water available to the soil, and therefore available for tree growth, depends on the amount of annual rainfall. Water may be lost by evaporation from the surface or by leaf transpiration. Evaporation and transpiration also control the temperature of the air in forests, which is always slightly warmer in cold months and cooler in warm months than the air in surrounding regions.

The sun sets on a savanna in the African country of Kenya.

The density of tree cover influences the amount of both sunlight and rainfall reaching every forest layer. A full-canopied forest absorbs between 60 and 90 percent of available light, most of which is absorbed by the leaves for photosynthesis. The movement of rainfall into the forest is considerably influenced by leaf cover, which tends to slow the velocity of falling water, which penetrates down to the ground level by running down tree trunks or dripping from leaves. Water not absorbed by the tree roots for nutrition runs along root channels, so water erosion is therefore not a major factor in shaping forest topography.

Forests are among the most complex ecosystems in the world, and they exhibit extensive vertical stratification. Conifer forests have the simplest structure: a tree layer rising to about 98 feet (30 metres), a shrub layer that is spotty or even absent, and a ground layer covered with lichens, mosses, and liverworts. Deciduous forests are more complex; the tree canopy is divided into an upper and lower story, while rainforest canopies are divided into at least three strata. The forest floor in both of these forests consists of a layer of organic matter overlying mineral soil. The humus layer of tropical soils is affected by the high levels of heat and humidity, which quickly decompose whatever organic matter exists. Fungi on the soil surface play an important role in the availability and distribution of nutrients, particularly in the northern coniferous forests. Some species of fungi live in partnership with the tree roots, while others are parasitically destructive.

Deciduous forest of beech in autumn, New Forest, southern England, U.K.

Nutrient cycling in tropical forests is dependent on fungi such as these growing in decaying leaf litter. Fungi decompose dead organic matter, releasing nutrients back into the ecosystem to be taken up by the roots of plants.

Animals that live in forests have highly developed hearing, and many are adapted for vertical movement through the environment. Because food other than ground plants is scarce, many ground-dwelling animals use forests only for shelter. In temperate forests, birds distribute plant seeds and insects aid in pollination, along with the wind. In tropical forests, fruit bats and birds effect pollination. The forest is nature's most efficient ecosystem, with a high rate of photosynthesis affecting both plant and animal systems in a series of complex organic relationships.

Temperate Forest Ecosystem

The temperate forest ecosystem is very important on Earth. Temperate forests are in regions where the climate changes a lot from summer to winter. Tropical rain forests are in regions where the climate stays constant all year long. Temperate forests are almost always made of two types of trees, deciduous and evergreen. Deciduous trees are trees that lose their leaves in the winter.

Evergreens are trees that keep them all year long, like pine trees. Forests can either be one or the other, or a combination of both. A fourth kind of forest is a temperate rain forest. These are found in California, Oregon and Washington in the United States.

These forests are made of redwoods and sequoias, the tallest trees in the world. The amount of rainfall in an area determines if a forest is present. If there is enough rain to support trees, then a forest will usually develop. Otherwise, the region will become grasslands.

Tropical Rain Forest Ecosystem

Tropical rain forests are one of the most important areas on Earth. These special ecosystems are homes to thousands of species animals and plants. Contrary to popular belief, rain forests are not only densely packed plants, but are also full of tall trees that form a ceiling from the Sun above. This ceiling keeps smaller plants from growing. Areas where sunlight can reach the surface are full of interesting plants.

Do you know where rain forests get their name? They are so named because they receive a lot of rain – an average of 80 inches a year. The temperature doesn't change very much during the year. It is always warm and muggy. The famous Amazon jungle is located in Brazil, in South America. This particular forest is called the Neotropics. Other large blocks are located in Central and West Africa.

Insects of the Tropical Rain Forest

The most feared and well known spider in the world resides in the jungle. Tarantulas are one of the creepiest animals you will ever see. Most species of tarantula have poisonous fangs for killing prey and for protection.

Although some are life-threatening to humans, others are harmless. Army ants are just one species of ant in the rain forest. They are called army ants because they march in a long, thick line through the jungle. They only stop when the young larvae reach pupil stage. Once the queen lays its eggs, the ants start marching again.

Beautiful butterflies fill the forest, but at one time these insects weren't so pretty. Butterflies start out as caterpillars, which tend to be a tad on the ugly side. They go through metamorphosis, which is the process of changing into a butterfly. Centipedes aren't so lucky. They don't turn into butterflies, but instead roam the forest looking for food. Some centipedes use poison to kill their prey.

Tropical Rain Forest Birds

The birds of the rain forest are the most beautiful in the world. A wide range of colors can be seen darting through the trees as the forest tops come to life. Many species of tropical birds are kept as pets because of their looks.

Hundreds of species of parrot live in the rain forest. The scarlet macaw is just one of these. It is also one of the longest, stretching to a length of 3 feet from its head to the tip of its tail. When these macaws eat a poisonous fruit, they eat a special type of clay that neutralizes the poison.

Toucans are also very interesting birds. They have large beaks that they use to reach fruit they can't get to. Scientists estimate there are 33 species of toucan in the rain forest. Not every tropical bird was blessed with looks. The hoatzin looks more like a peacock without the pretty tail.

Hoatzins are terrible flyers crash landings are common practice. The brown kiwi is a flightless bird that looks more like a rodent with a long beak and feathers. Kiwis live on the ground instead of the trees. They have special claws used for running, digging and defence.

Tropical Rain Forest Mammals

Birds aren't the only creatures that fly through the rain forests. Several species of flying mammals live in the jungle. From the harmless fruit bat to the unique flying squirrel, the tropical rain forests are full of surprises.

The Indian flying fox is one of the largest bats in the world. Its wings can spread out to 5 feet in width. Unlike bats in other parts of the world, these bats do not live in caves. They prefer to.hang in trees during the day. Hundreds or even thousands of bats can be spotted in a single tree.

Vampire bats live in the Amazon jungle in South America. The famous stories of blood-sucking bats probably originated here. These bats do in fact drink the blood of their victims. They usually attack farm animals, but have also enjoyed the blood of humans. But vampire bats only drink a very small amount of fluid.

Tropical Rain Forest Reptiles

The tropical rain forests of the world are full of reptiles. Reptiles are cold blooded, which means their body temperature depends on their environment. So, it is important for them to stay in warm climates. Snakes are reptiles, and the rain forests are home to many. The mamba family is the most poisonous of all. They kill their prey by injecting poison with their sharp fangs.

Anacondas make up another snake family. They are some of the longest creatures in the world, as they can reach 30 feet in length. Anacondas prefer to wrap themselves around their prey and squeeze, rather than inject poison. Anacondas swallow their prey whole and sleep while the food is digesting. Chameleons are interesting lizards that can change color.

This enables them to blend in with their surroundings. Not only is this a great disguise from predators, it is also an easy way to sneak up on their prey. Chameleons only eat insects. Geckos are very neat creatures. The flying gecko can glide from tree to tree to escape from predators. Their grip is so strong, that if you tried to pull one off a window, the glass would break before the gecko would let go.

Tropical Rain Forest Primates

Monkeys and their cousins are all primates. Humans are also primates. There are many species of monkeys in the tropical rain forests of the world. Monkeys can be divided into two groups: new world monkeys and old world monkeys. New world monkeys live only in South and Central America. Spider monkeys live in the rain forests in the Andes Mountains.

They look very strange with their long noses. Spider monkeys eat mostly fruit and nuts, so they are called frugivores. They are joined by the howler monkeys. These primates are so named because they have a special sac that makes their sounds louder.

Old world monkeys live only in Africa and Asia. The colobus monkey is one such kind. These monkeys are called foliovores because they eat leaves. They live in small groups of 15, but other primates live in larger groups of up to 200. There are too many species. Chimpanzees, orangutans and gorillas are all called pongids. These primates are more famous than the others. Gorillas are too big to climb trees, so they are found on the forest floor.

Boreal or Taiga Forests

The boreal forest ecosystem is the contiguous green belt of conifer and deciduous trees that encircles a large portion of the Northern Hemisphere. In North America, the boreal forest stretches across most of northern Canada and into Alaska. It has long been identified as one of the world's great forest ecosystems.

This forest ecosystem covers roughly 35% of Canada's land mass and is the single largest land based ecosystem in North America. It also contains a significant proportion of Canada's biodiversity and has long been recognized as an important global carbon sink.

Although the boreal is relatively unknown, it is important as the "great lung" of North America, "breathing in" carbon dioxide and "exhaling" oxygen into the atmosphere. In short, the boreal forest manages to do what the rain forest of the Amazon does but with only the fraction of the flora and fauna.

This forest ecosystem houses the largest and smallest mammal species (wood bison & pygmy shrews) of the North American continent. The Boreal forest has many things: great lakes and northern rivers; vast bogs, fens and other organic wetlands. The rich wildlife diversity of the Boreal is a joy to behold: woodland caribou and lynx; whooping cranes and wood bison; northern owls; woodpeckers with three rather than four toes; colorful wood warblers.

The Boreal has more than 5,000 species of conspicuous and colorful fungi, celebrated far more in Scandinavia and Siberia than in North America. Then there are the precious old-growth forests, the richest and most biologically diverse of the Boreal forest communities that are essential for so many Boreal species.

Structure of Forest Ecosystems

Different organisms exist within the forest layers. These organisms interact with each other and their surroundings. Each organism has a role or niche in sustaining the ecosystem.

Some provide food for other organisms; others provide shelter or control populations through predation.

Producers

All living organisms intake energy in order to survive. In a forest ecosystem, trees and other plants get their energy from sunlight. Plants produce their own food, in the form of carbohydrates. Plants are, therefore, called the primary producers, since they produce the basic foodstuffs for other organisms within food chains and food webs. Photosynthesis is the chemical reaction that allows plants to produce their own food.

Consumers

Animals cannot produce their own food. They must consume food sources for die energy they need to survive. All animals, including mammals, insects, and birds, are called consumers. Consumers rely on plants and other animals as a food source.

Primary consumers only eat plants and are referred to as herbivores. Secondary consumers are referred to as carnivores and feed on herbivores. Tertiary consumers are carnivores that feed on other carnivores. Omnivores eat both plant and animal matter.

Decomposers

Leaves, needles, and old branches fall to the forest floor as trees grow. Eventually all plants and animals die. These materials are decomposed by worms, microbes, fungi, ants, and other bugs.

Decomposers break these items down into their smallest primary elements to be used again. Decomposers are important in that they sustain the nutrient cycle of ecosystems.

Humans are Part of Forest Ecosystem

Humans are consumers. We get food and materials from forests. Because of this, we are a part of the forest ecosystem. Human consumption alters forest ecosystems. Human intervention may be necessary to sustain forest communities under the increased pressure of human use.

Grassland Ecosystem

A grassland ecosystem is a community of creatures living together within a grassy space. These creatures can include various types of grasses, insects, and animals, etc. Grassland is an area in which the vegetation is dominated by a nearly continuous cover of grasses. Grasslands occur in environments conducive to the growth of this plant cover but not to that of taller plants, particularly trees and shrubs. The factors preventing establishment of such taller, woody vegetation are varied.

Grasslands are one of the most widespread of all the major vegetation types of the world. This is so, however, only because human manipulation of the land has significantly altered the natural vegetation, creating artificial grasslands of cereal crops, pastures, and other areas that require some form of repetitious, unnatural disturbance such as cultivation, heavy grazing, burning, or mowing to persist.

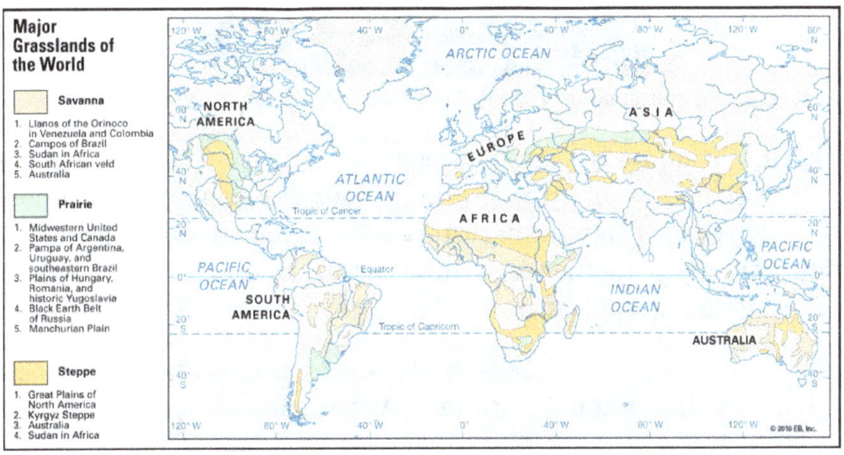

Grassland regions: savanna, prairie, and steppe.
Principal regions where significant areas of natural savanna, prairie, and steppe occur.

The most extensive natural grasslands can be thought of as intermediates in an environmental gradient, with forests at one end and deserts at the other. Forests occupy the most favourable environments, where moisture is adequate for growth and survival of a tall, dense vegetation dominated by trees. Deserts are found where moisture is so lacking that a continuous, permanent vegetation cover cannot be maintained. Grasslands lie between these two extremes.

Like the savannas, deserts, and scrublands into which they commonly blend, grasslands arose during the period of cooling and drying of the global climate, which occurred during the Cenozoic Era (65.5 million years ago to the present). Indeed, the grass family itself (Poaceae or Gramineae) evolved only early in this era. The date of earliest appearance of grasslands varies from region to region. In several regions a succession of vegetation types can be recognized in the Cenozoic fossil record, as climate dried out progressively. For example, in central Australia during the past 50 million years tropical rainforest gave way successively to savanna, grassland, and, finally, desert. In some places expansion of grasslands to something approaching their modern extent occurred only during the extremely cold, dry intervals called ice ages in north temperate regions of the past two million years.

Highveld grassland near Heidelberg, S.Af., southeast of Johannesburg.

A dynamic balance commonly exists between grasslands and related vegetation types. Droughts, fires, or episodes of heavy grazing favour grassland at some times, and wet seasons and an absence of significant disturbances favour woody vegetation at others. Changes in the severity or frequency of these factors can cause a change from one vegetation type to another.

Other grassland types occur in places too cold for trees to grow i.e., beyond the forest limits of high mountains or at high latitudes. A characteristic type of grassland in cool, moist parts of the Southern Hemisphere is tussock grassland, dominated by tussock or bunch grasses that develop pedestals of matted stems, giving the vegetation a lumpy appearance. Tussock grasslands occur at various latitudes. In the tropics they are found above the forest limit on some high mountains e.g., in New Guinea and East Africa. At the higher latitudes of the Southern Ocean they form the main vegetation of subantarctic islands. They are also typical of the drier, colder parts of New Zealand and the southernmost regions of South America.

Not all natural grasslands, however, arise from climate-related circumstances. Woody plants may be prevented from growing in certain areas for other reasons, allowing grasses to dominate. One cause is seasonal flooding or waterlogging, which is responsible for the creation and maintenance of large grasslands in parts of the highly seasonal subtropics and in smaller areas of other regions. One of the best examples of a seasonally flooded subtropical grassland is the Pantanal in the Mato Grosso region of Brazil. Across an area of 140,000 square kilometres (54,000 square miles), dry grasslands prevail for half of each year and shallow wetlands for the other, with small forest patches restricted to low rises that do not flood during the wet season. In many other areas where climate is suitable for forest growth, very shallow or infertile soils may prevent tree growth and result in development of grassland.

The largest areas of natural grassland those resulting from climatic dryness can be classified into two broad categories: tropical grasslands, which generally lie between the belts of tropical forest and desert; and temperate grasslands, which generally lie between deserts and temperate forests. Tropical grasslands occur in the same regions as savannas, and the distinction between these two vegetation types is rather arbitrary, depending on whether there are few or many trees. Likewise, temperate grasslands may have a scattering of shrubs or trees that blurs their boundaries when they occur adjacent to scrublands or temperate forests.

Tropical grasslands are found mainly in the Sahel south of the Sahara, in East Africa, and in Australia. Temperate grasslands principally occur in North America, Argentina, and across a broad band from Ukraine to China, but in most of these regions they have been substantially altered by agricultural activities.

Many grasslands formerly supposed to be natural are now recognized as having once been forests that grew in a marginally dry climate. Early human disturbance is responsible for their transformation. For example, almost the entire extensive lowland grasslands of the eastern part of the South Island, New Zealand, are believed to have been created by forest-burning carried out by the Polynesians the country's first colonists during the eight centuries before European settlement in the 18th century.

Seminatural grasslands may occur where woody vegetation was once cleared for agricultural purposes that have since been abandoned; a return to the original vegetation is prevented by repeated burning or grazing. In wet tropical regions these types of grasslands may be very dense, such as those in East Africa that are dominated by elephant grass (Pennisetum purpureum) or in New Guinea by pit-pit grass (Miscanthus floridulus), both of which grow 3 metres (9.8 feet) tall.

All areas of grassland may owe something of their area and character to a long history of interaction with humans, particularly through the medium of fire.

Environment

Grassland climates are varied, but all large regions of natural grassland are generally hot, at least in summer, and dry, though not to the extent that deserts are. In general, tropical grasslands receive 500 to 1,500 millimetres (20 to 60 inches) of rain in an average year and in every season experience temperatures of about 15 to 35 °C (59 to 95 °F). The dry season may last as long as eight months. An excess of rainfall over evaporation, leading to ephemeral river flow, occurs only during the wet season. The tropical grassland climate overlaps very broadly with that of savanna. As previously stated, these vegetation types differ little from each other, a savanna being merely a grassland with scattered trees. Small changes in management and usage can convert one to the other.

Temperate grasslands are somewhat drier than tropical grasslands and also colder, at least for part of the year. Seasonal temperature variation may be slight in tropical grasslands but may vary by as much as 40 °C (72 °F) in temperate grassland areas. Mean annual rainfall in the North American grassland areas is 300 to 600 millimetres. Mean temperatures in January range from 18 °C (0 °F) in the north to 10 °C (50 °F) in the south, with corresponding values in July being 18 °C (64 °F) and 28 °C (82 °F). Mean annual temperature in the most northerly areas of the North American grassland zone is below 0 °C (32 °F).

Occurring as they do across a wide range of climatic and geologic conditions, grasslands are associated with many different types of soil. The grassland ecosystem itself influences soil formation, and this causes grassland soils to differ from other soils. The nature of grass litter and its pattern of decomposition commonly result in the development of a dark, organically rich upper soil layer that can reach 300 millimetres below the surface. This layer is absent from desert soils and is different from the surface layer of rotting leaf litter typical of forest soils. It is friable in structure and rich in plant nutrients. Lower soil layers are typically pale and yellowish, especially at depths close to two metres.

Biota

In pre-European North America, grasslands spread across a large portion of the continent, from the Rocky Mountains in the west to the deciduous forests in the east. Of this vast expanse, only tiny fragments remain in any condition remotely similar to their original state. The largest central area consisted of mixed prairie, dominated by several species of the grasses Stipa, Agropyron, Bouteloua, and Koeleria. Mixed prairie gave way in the north to a fescue prairie with Festuca and Helictotrichon; in the west, to a short-grass steppe dominated by Bouteloua gracilis and Buchloe dactyloides; and to the east, to a tall-grass prairie with the bluestem grasses Andropogon gerardii and A. scoparium. Trees and shrubs were generally absent, but a large variety of herbaceous plants occurred with the grasses.

The large grazing mammals of the North American prairies included the bison and pronghorn antelope, whose typical predator was the gray wolf. The badger and several rabbit and hare species were widespread, as were many small burrowing rodents. Among the invertebrate fauna, grasshoppers were and still are particularly important. In grasslands the total biomass of invertebrates typically exceeds that of the much larger and more conspicuous vertebrates, except in regions containing large numbers of domestic stock.

California grassland.
A California grassland at dusk.

The principal region of grassland in South America is in the southeast portion of the continent; it can be divided into the Pampas of Argentina and the campos of the adjacent areas of Uruguay and Brazil. Among the many grasses in the Pampas, Stipa is the most diverse, while another suite of grasses in the campos includes among its more common members Paspalum and Andropogon. The vegetation of the region is now greatly altered by centuries of heavy grazing and burning. Before this, the principal large herbivore was the pampas deer, of which only one small herd survives. Opossums, armadillos, and rodents also were abundant and were preyed upon by various cats, such as jaguars and pumas, and foxes. The rhea, a large, flightless bird, is indigenous to this area.

The Eurasian steppe, occupying an extensive tract of the former Soviet Union and Mongolia, is in many ways very similar to the prairie of North America and hence harbours many similar plants and animals. Various species of Stipa dominate the flora in most areas, and they are mixed in spots with other grasses, of which Festuca and Agropyron assume dominance. Through their feeding and burrowing activities rodents are important to the maintenance and composition of the vegetation; species include the large marmots and a diversity of voles and other smaller types. A vole in Mongolia, Lasiopodomys brandtii, in some years can consume such a high proportion of the vegetation that it reduces its grassland habitat to virtual desert.

The Sahel the broad band of grassland crossing western and north-central Africa south of the Sahara is the largest area of tropical grassland. For millennia human populations have put many demands on the region, so that its present condition is quite unlike its natural condition. The most common grasses include Aristida, Cenchrus, and Schoenefeldia. Other species, which are highly palatable to grazing animals, are now restricted to rocky sites that offer some protection; these species may have once been far more widespread and important. In many places where shrubs and small trees occur the vegetation would be called more accurately a scrubland or savanna were it not so easily transformed into grassland by practices such as grazing, burning, and fuel gathering.

The grasslands of East Africa include wetter environments than exist in the Sahel and hence are more diverse. Where forests have been destroyed, a tall grassland consisting of Pennisetum or Hyparrhenia develops and may be kept in this condition indefinitely through burning or through the browsing and grazing of such herbivores as elephants. Other grasses such as Aristida and Chrysopogon are important in drier sites, and Themeda occurs in cooler places at higher altitudes. Herbivorous mammals include wildebeests, several antelope species, and—where they still survive rhinoceroses, buffalo, and elephants. Carnivores include various dogs (jackals), cats (cheetahs, lions), hyenas, and mongooses.

Tropical grasslands in Australia in the extensive arid areas are generally dominated by species of the spinifex grasses, Plectrachne and Triodia, which form characteristic hummocks by trapping windblown sand at the bases of their tussocks. Heteropogon and Sorghum dominate grasslands in moister, northern areas, and Astrebla (Mitchell grass) is prevalent in seasonally arid areas, especially on cracking clay soils in the east. Other grass species are usually subordinate but may dominate in spots. Woody plants, particularly Acacia in arid areas and Eucalyptus in moister places, may be so numerous that the vegetation cannot be considered true grassland. The Mitchell grasslands were once much purer until they were altered by heavy grazing of domestic stock; today, vast tracts have been invaded by the African shrub Acacia nilotica, introduced by humans.

The largest native animals in Australian grasslands are kangaroos, of which the biggest is the red kangaroo, characteristic of the dry inland areas where natural grasslands are found. Mammals introduced from other continents, however, have become as common; these include domesticated stock, especially cattle and sheep, and a range of feral fauna such as camels, horses, donkeys, and goats. European rabbits are also widespread, abundant, and highly destructive. A main predator is the dingo, or wild dog. Reptiles, especially lizards, are very diverse; birds include the large flightless emu as well as a wide range of parrots and other flying forms.

The tussock grasslands of New Zealand and the subantarctic islands are commonly dominated by species of Poa. Related vegetation also occurs on high mountains in equatorial regions. These grasses have in most cases evolved in the absence of grazing mammals; the introduction of exotic fauna such as deer and sheep has caused severe degradation of the vegetation in many places.

Population and Community Development and Structure

Whether tropical or temperate, natural grasslands occur in environments in which growing conditions are favourable for only a short season. In tropical regions this growing season is usually the rainy season or, in some cases, the season when the ground is not waterlogged or submerged. In temperate grasslands the growing season is usually the short period between the cold, damp winter and the hot, dry summer. Perennial grasses, relying on subterranean reserves of stored food for rapid shoot growth, are well adapted to exploiting such brief growing seasons, reaching their maximum size and completing their seeding within a few weeks. Their aboveground parts then die back, providing potential fuel for the grass fires that typify these environments. The underground perennating roots and rhizomes of the grasses, however, are relatively well protected from fire.

Vegetation profile of a typical grassland.

Grasslands tend to produce larger amounts of new growth if subjected to some type of repeated disturbance, usually grazing or fire, that prevents the accumulation of a thick layer of dead litter. Where such a layer is allowed to develop, it retains nutrients in a form not immediately available to roots and acts as a physical barrier for new shoots growing from the soil surface toward the light; in temperate grasslands this layer acts as thermal insulation, slowing the spring warming of the soil. This has obvious implications for grazing management of these systems.

Woody plants, which might be expected to shade the grasses and dominate the vegetation, are disadvantaged by the shortness of the growing season. Nevertheless, in the absence of heavy mammalian grazing and especially of regular fires, some trees and shrubs that grow vigorously may become established. Thus, the grasslands in such situations are maintained by these natural, or seminatural, disturbances of fire and grazing, which prevent the succession of the grassland vegetation toward tropical deciduous forest, savanna, scrubland, or temperate forest.

Biological Productivity

Because of its importance for grazing and other grassland agricultural production, grassland productivity has been extensively investigated using various methods. However, most studies have focused only on aboveground productivity, ignoring the important subterranean component, which can be much more substantial as much as 10 times greater even when the aboveground portion is at a seasonal maximum. Typical aboveground biomass (dry weight of organic matter in an area) values for North American grasslands are 2.5 to 6 metric tons per hectare, of which about three-quarters is in the form of dead shoot material; however, values up to and greater than 20 metric tons per hectare have been recorded in some tropical grasslands. (For a full discussion of productivity see biosphere: The organism and the environment: Resources of the biosphere).

There is a relatively rapid turnover of plant matter in grasslands. Most plant organs have a life span of only one or a few seasons, leading to annual rates of overall turnover of about 20 to 50 percent. When consumption of plant parts by consumers is taken into account, annual productivity values are similar to biomass values. Aboveground annual productivity in a typical Canadian grassland was found to be 6.4 metric tons per hectare when all leaching and decomposition losses from dead shoots were taken into consideration. Values found in other grasslands have varied greatly but often have been significantly lower than this, partly because such losses have not been fully taken into account. Repeated harvesting (surrogate grazing) commonly yields 1.5 to 2.5 metric tons per hectare per year.

Grasslands frequently have been converted to cropland on which edible grains are grown; this allows food for humans to be taken directly from the grasslands themselves rather than via grazing animals feeding on the native grasses in a rangeland situation. The increase in yields is substantial. According to figures compiled for the northern Great Plains of North America, for example, one hectare of natural grassland grazed by cattle can provide the equivalent of one-tenth of one person's annual food energy requirements, whereas the same area sown to wheat can provide four persons' requirements.

However, it should be pointed out that wheat yields cannot be sustained annually at their initially high levels without additional inputs of fertilizer. Similarly, sustained heavy grazing of natural grasslands can quickly lead to degradation of the vegetation and erosion of the soil. This is particularly likely to occur unnoticed in regions that lack fences and therefore do not exhibit fenceline contrasts in range condition. In these fenceless grazing regions degradation due to overgrazing may be difficult to distinguish from background changes due to long or short-term weather variations.

Desert Ecosystem

Desert ecology is the study of interactions between both biotic and abiotic components of desert environments. A desert ecosystem is defined by interactions between organism populations, the climate in which they live, and any other non-living influences on the habitat. Deserts are arid regions which are generally associated with warm temperatures, however cold deserts also exist. Deserts can be found in every continent, with the largest deserts located in Antarctica, the Arctic, Northern Africa, and the Middle East.

Sand dunes in the Sahara Desert.

Climate

Temperatures were 115° high enough to melt the street signs.

Deserts experience a wide range of temperatures and weather conditions, and can be classified into four types: hot, semiarid, coastal, and cold. Hot deserts experience warm temperatures year round, and low annual precipitation. Low levels of humidity in hot deserts contribute to high daytime temperatures, and extensive nighttime heat loss. The average annual temperature in hot deserts is approximately 20 to 25 °C, however, extreme weather conditions can lead to temperatures ranging from -18 to 49 °C.

Rainfall generally occurs in concentrated bursts, followed by long periods of dryness. Semiarid deserts experience similar conditions to hot deserts, however, the maximum and minimum temperatures tend to be less extreme, and generally range from 10 to 38 °C. Coastal deserts are cooler than hot and semiarid deserts, with average summer temperatures ranging between 13 and 24 °C. They also feature higher total rainfall values. Cold deserts are similar in temperature to coastal deserts, however, they receive more annual precipitation in the form of snowfall. Deserts are most notable for their dry climates; usually a result from their surrounding geography. For example, rain-blocking mountain ranges, and distance from oceans are two geographic features that contribute to desert aridity. Rain-blocking mountain ranges create Rain Shadows. As air rises and cools, its relative humidity increases and some or most moisture rains out, leaving little to no water vapor to form precipitation on the other side of the mountain range.

Deserts occupy one-fifth of the Earth's land surface and occur in two belts: between 15° and 35° latitude in both the southern and northern hemispheres. These bands are associated with the high solar intensities that all areas in the tropics receive, and with the dry air brought down by the descending arms of both the Hadley and Ferell atmospheric circulation cells. Dry winds hold little moisture for these areas, and also tend to evaporate any water present.

Many desert ecosystems are limited by available water levels, rather than rates of radiation or temperature. Water flow in these ecosystems can be thought of as similar to energy flow; in fact, it is often useful to look at water and energy flow together when studying desert ecosystems and ecology.

Water availability in deserts may also be hindered by loose sediments. Dust clouds commonly form in windy, arid climates. Scientists have previously theorised that desert dust clouds would enhance rainfall, however, some more recent studies have shown that precipitation is actually inhibited by this phenomenon by absorbing moisture from the atmosphere. This absorption of atmospheric moisture can result in a positive feedback loop, which leads to further desertification.

Landscape

Erosion over time due to wind and rains have created beautiful canyons in the landscape. Antelope Canyon, AZ.

Desert landscapes can contain a wide variety of geological features, such as oases, rock outcrops, dunes, and mountains. Dunes are structures formed by wind moving sediments into mounds. Desert dunes are generally classified based on their orientation relative to wind directly. Possibly the most recognizable dune type are transverse dunes, characterized by crests transverse to the wind direction. Many dunes are considered to be active, meaning that they can travel and change over time due to the influence of the wind. However, some dunes can be anchored in place by vegetation or topography, preventing their movement. Some dunes may also be referred to as sticky.

These types of dunes occur when individual grains of sand become cemented together. Sticky dunes tend to be more stable, and resistant to wind reworking than loose dunes. Barchan, and Seif dunes are among the most common of desert dunes. Barchan dunes are formed as winds continuously blow in the same direction, and are characterized by a crescent-shape atop the dune. Seif dunes are long and narrow, featuring a sharp crest, and are more common in the Sahara Desert.

Analysis of geological features in desert environments can reveal a lot about the geologic history of the area. Through observation and identification of rock deposits, geologists are able to interpret the order of events that occurred during desert formation. For example, research conducted on the surface geology of the Namib Desert allowed geologists to interpret ancient movements of the Kuiseb River based on rock ages and features identified in the area.

Organism Adaptation

The high body temperatures of bactrian camels allow them to preserve water.

Animals

Deserts support diverse communities of plant and animals that have evolved resistance, and circumventing methods of extreme temperatures and arid conditions. For example, desert grasslands are more humid and slightly cooler than its surrounding ecosystems. Many animals obtain energy by eating the surrounding vegetation, however, desert plants are much more difficult for organisms to consume. To avoid intense temperatures, the majority of small desert mammals are nocturnal, living in burrows to avoid the intense desert sun during the daytime. These burrows prevent overheating and dehydration as they maintain an optimal temperature for the mammal. Desert ecology is characterized by dry, alkaline soils, low net production and opportunistic feeding patterns by herbivores and carnivores. Other organisms' survival tactics are physiologically based. Such tactics include the completion of life cycles ahead of anticipated drought seasons, and storing water with the help of specialized organs.

The hairy, white bristles of the old man cactus help deflect strong desert sunlight.

Desert climates are particularly demanding on endothermic organisms. However, endothermic organisms have adapted mechanisms to aid in water retention in habitats such as desert ecosystems which are commonly affected by drought. In environments where the external temperature is less than their body temperature, most endotherms are able to balance heat production and heat loss to maintain a comfortable temperature. However, in deserts where air and ground temperatures exceed body temperature, endotherms must be able to dissipate the large amounts of heat being absorbed in these environments. In order to cope with extreme conditions, desert endotherms have adapted through the means of avoidance, relaxation of homeostasis, and specializations. Nocturnal desert rodents, like the kangaroo rat, will spend the daytime in cool burrows deep underground, and emerge at night to seek food. Birds are much more mobile than ground-dwelling endotherms, and can therefore avoid heat-induced dehydration by flying between water sources. To prevent overheating, the body temperatures of many desert mammals have adapted to be much higher than non-desert mammals. Camels, for example, can maintain body temperatures that are about equal to typical desert air temperatures. This adaptations allows camels to retain large amounts of water for extended periods of time. Other examples of higher body temperature in desert mammals include the diurnal antelope ground squirrel, and the oryx. Certain desert endotherms have evolved very specific and unique characteristics to combat dehydration. Male sandgrouse have specialized belly feathers that are able to trap and carry water. This allows the sandgrouse to provide a source of hydration for their chicks, who do not yet have the ability to fly to water sources themselves.

Plants

Although deserts have severe climates, some plants still manage to grow. Plants that can survive in arid deserts are called xerophytes, meaning they are able to survive long dry periods. Such plants may close their stomata during the daytime and open them again at night. During the night, temperatures are much cooler, and plants will experience less water loss, and intake larger amounts of carbon dioxide for photosynthesis.

Adaptations in xerophytes include resistance to heat and water loss, increased water storage capabilities, and reduced surface area of leaves. One of the most common families of desert plants are the cacti, which are covered in sharp spines or bristles for defence against herbivory. The bristles on certain cacti also have the ability to reflect sunlight, such as those of the old man cactus. Certain xerophytes, like oleander, feature stomata that are recessed as a form of protection against hot, dry desert winds, which allows the leaves to retain water more effectively. Another unique adaptation can be found in xerophytes like ocotillo, which are "leafless during most of the year, thereby avoiding excessive water loss".

There are also plants called phreatophytes which have adapted to the harsh desert conditions by developing extremely long root systems, some of which are 80 ft. long; to reach the water table which ensures a water supply to the plant.

Types and Characteristic Features

One can find at least one desert on every continent except Europe and Antarctica. Each desert is different in some way, but they all have one thing in common. In order for an area of land to be considered a desert, it must receive less than 10 inches of water a year.

How come deserts get such little water? Clouds are scarce in these regions, and we all know that without clouds, there can't be rain, snow or any other precipitation. But clouds also serve another purpose they block out some of the Sun. The desert gets mighty hot during the day because the Sun beats down on the sand. At night, the desert gets very cold, because there aren't clouds around to keep the heat from escaping to the atmosphere.

There are plenty of differences between the deserts of the world. Some deserts are made of very fine, red sand, others consist of sand mixed with pebbles and rocks. The desert sand started out as rock, but years of weathering by wind and water has created dunes in the deserts. These sands are mostly minerals, and sometimes oil can be found hidden deep within the rocks.

Structure and Function

The different components of a desert ecosystem are:

Abiotic Component

The abiotic component includes the nutrients present in the soil and the aerial environment. The characteristic feature of the abiotic component is lack of organic matter in the soil and scarcity of water.

Biotic Component

The various biotic components representing three functional groups are:

Producer Organisms

The producers are mainly shrubs or bushes, some grasses and a few trees. Surprisingly, there are many species of plants that survive in the desert. Most of them are succulents, which mean they store water. Others have seeds that lay dormant until a rain awakens them. Regardless, these plants find a way to get water and protect themselves from the heat.

The most famous desert plant is the cactus. There are many species of cacti. The saguaro cactus is the tall, pole shaped cactus. The saguaro can grow up to 40 feet tall. It can hold several tons of water inside its soft tissue. Like all cacti, the saguaro has a thick, waxy layer that protects it from the Sun.

Other succulents include the desert rose and the living rock. This strange plant looks like a spiny rock. It's disguise protects it from predators. The welwitschia is a weird looking plant. It has two long leaves and a big root. This plant is actually a type of tree and it can live for thousands of years.

There are many other kinds of desert plants. Some of them have thorns others have beautiful flowers and deadly poisons. Even in the worst conditions, these plants continue to thrive.

Consumers

These include animals such as insects and reptiles. Besides them, some rodents, birds and some mammalian vertebrates are also found.

Desert Insects and Arachnids

There are plenty of insects in the desert. One of the most common and destructive pests is the locust. A locust is a special type of grasshopper. They travel from place to place, eating all the vegetation they find. Locusts can destroy many crops in a single day.

Not all desert insects are bad, though. The yucca moth is very important to the yucca plant, because it carries pollen from the flower to the stigma. The darkling beetle has a hard, white, wing case that reflects the Sun's energy. This allows the bug to look for food during the day.

There are also several species of ants in the desert. The harvester ants gather seeds and store them for use during the dry season. And the honey pot ants have a very weird habit. Some members of the colony eat large amounts of sugar, so much that their abdomens get too large for them to move. The rest of the colony feeds off this sugar.

There are also arachnids in the desert. Spiders are the most notable arachnids, but scorpions also belong in this group. Some species of scorpions have poison in their sharp tails. They sting their predators and their prey with the piercing tip.

Desert Reptiles

Reptiles are some of the most interesting creatures of the desert. Reptiles can withstand the extreme temperatures because they can control their body temperatures very easily. You can put most of the desert reptiles into one of two categories: snakes and lizards.

Many species of rattlesnakes can be found in the desert. Rattlesnakes have a noisy rattle they use to warn enemies to stay away. If the predator isn't careful, the rattlesnake will strike, injecting venom with its sharp fangs. Other desert snakes include the cobra, king snake and the hognose.

Lizards make up the second category of desert reptiles. They are probably the most bizarre looking animals in the desert. While some change colors and have sharp scales for defense, others change their appearance to look more threatening.

One such creature is the frilled hazard. When enemies are near, the lizard opens its mouth, unveiling a wide frill. This makes the hazard look bigger and scarier. The shingle back has a tail with the same shape as its head. When a predator bites at the tail, the shingle back turns around and bites back. There are only two venomous lizards in the world, and one of them is the gila monster. It has a very painful bite.

Desert Birds

Like the other inhabitants of the desert, birds come up with interesting ways to survive in the harsh climate. The sand grouse has special feathers that soak up water. It can then carry the water to its young trapped in the nest.

Other birds, like the gila woodpecker, depend on the giant saguaro as its home. This woodpecker hollows out a hole in the cactus for a nest. The cool, damp inside is safe for the babies.

The roadrunner is probably the most well known desert bird. Roadrunners are so named because they prefer to run rather than fly. Ostriches also prefer to use their feet. Even the young depend on walking to find food and water. The galah is one of the prettiest desert birds. It is one of the few species that return to the same nest year after year.

Galahs are interesting birds, in that the number of eggs they lay depends on the climate. If the desert is in a drought, they don't lay any. However, during more tolerable years, the galah may lay as many as five eggs.

Desert Mammals

There are several species of mammals in the desert. They range in size from a few inches to several feet in length. Like other desert wildlife, mammals have to find ways to stay cool and drink plenty of water. Many desert mammals are burrowers.

They dig holes in the ground and stay there during the hot days. They return to the surface at night to feed. Hamsters, rats and their relatives are all burrowers. Not only do the burrows keep the animals cool, they are also a great place to store food.

Of course, not all animals have in holes in the ground. The kangaroo and spiny anteater both live in the Australian desert region. Spiny anteaters are unusual mammals because they lay eggs.

The desert is also full of wild horses, foxes and jackals, which are part of the canine family. And we can't forget the cats. Lions are found all over the deserts of southern Africa. They get their water from the blood of their prey.

Camels

Camels could be included in the mammal section. Camels are the cars of the desert. Without them, people would have great difficulty crossing the hot terrain. There are two types of camels: Bactrian and dromedary. The main difference between the two is the number of humps. Dromedaries have one hump, and Bactrian have two. Both kinds are used by people, but only Bactrian's are found in the wild.

Camels are great for transportation because they use very little water. Camels can withstand very high temperatures without sweating. They also store fat in their humps for food. If a Bactrian camel travels a long distance without eating, its hump will actually get smaller.

Decomposers

Due to poor vegetation the amount of dead organic matter is very less. As a result the decomposers are very few. The common decomposers are some bacteria and fungi, most of which are thermophile.

Aquatic Ecosystem

An aquatic ecosystem is an ecosystem in a body of water. Communities of organisms that are dependent on each other and on their environment live in aquatic ecosystems. The two main types of aquatic ecosystems are marine ecosystems and freshwater ecosystems.

An estuary mouth and coastal waters, part of an aquatic ecosystem.

Types

Marine

Marine ecosystems, the largest of all ecosystems, cover approximately 71% of the Earth's surface and contain approximately 97% of the planet's water. They generate 32% of the world's net primary production. They are distinguished from freshwater ecosystems by the presence of dissolved compounds, especially salts, in the water. Approximately 85% of the dissolved materials in seawater are sodium and chlorine. Seawater has an average salinity of 35 parts per thousand of water. Actual salinity varies among different marine ecosystems.

Marine ecosystems can be divided into many zones depending upon water depth and shoreline features. The oceanic zone is the vast open part of the ocean where animals such as whales, sharks, and tuna live. The benthic zone consists of substrates below water where many invertebrates live. The intertidal zone is the area between high and low tides; in this figure it is termed the littoral zone. Other near-shore (neritic) zones can include estuaries, salt marshes, coral reefs, lagoons and mangrove swamps. In the deep water, hydrothermal vents may occur where chemosynthetic sulfur bacteria form the base of the food web.

Classes of organisms found in marine ecosystems include brown algae, dinoflagellates, corals, cephalopods, echinoderms, and sharks. Fishes caught in marine ecosystems are the biggest source of commercial foods obtained from wild populations.

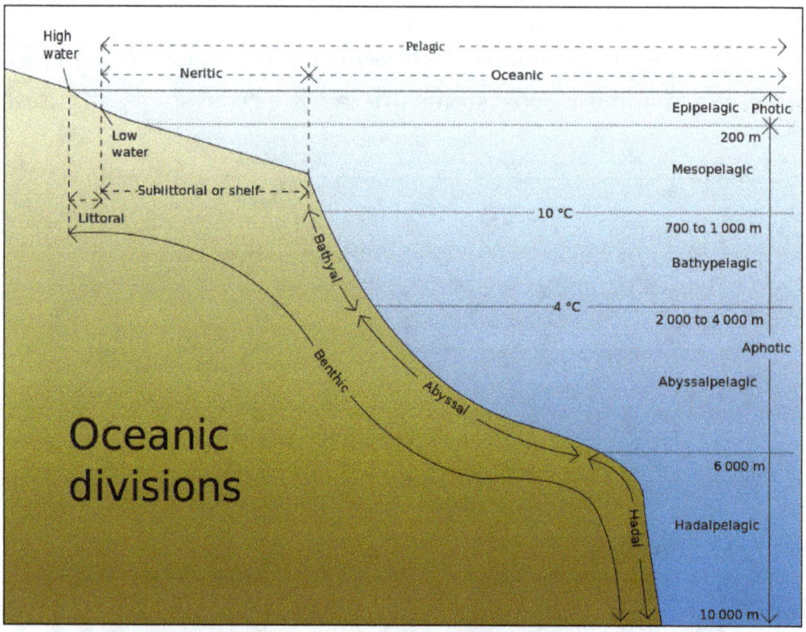

A classification of marine habitats.

Environmental problems concerning marine ecosystems include unsustainable exploitation of marine resources (for example overfishing of certain species), marine pollution, climate change, and building on coastal areas.

Freshwater

Freshwater ecosystem.

Freshwater ecosystems cover 0.78% of the Earth's surface and inhabit 0.009% of its total water. They generate nearly 3% of its net primary production. Freshwater ecosystems contain 41% of the world's known fish species.

There are three basic types of freshwater ecosystems:

- Lentic: Slow moving water, including pools, ponds, and lakes.
- Lotic: Faster moving water, for example streams and rivers.

- Wetlands: Areas where the soil is saturated or inundated for at least part of the time.

Lentic

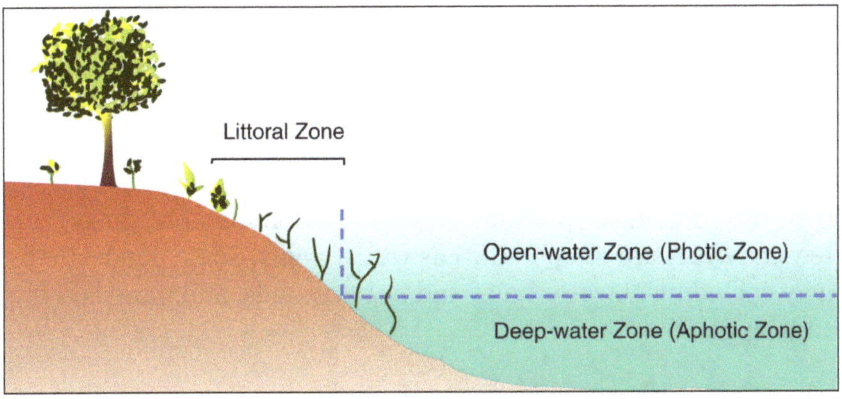

The three primary zones of a lake.

Lake ecosystems can be divided into zones. One common system divides lakes into three zones. The first, the littoral zone, is the shallow zone near the shore. This is where rooted wetland plants occur. The offshore is divided into two further zones, an open water zone and a deep water zone. In the open water zone (or photic zone) sunlight supports photosynthetic algae, and the species that feed upon them. In the deep water zone, sunlight is not available and the food web is based on detritus entering from the littoral and photic zones. Some systems use other names. The off shore areas may be called the pelagic zone, the photic zone may be called the limnetic zone and the aphotic zone may be called the profundal zone. Inland from the littoral zone one can also frequently identify a riparian zone which has plants still affected by the presence of the lake this can include effects from windfalls, spring flooding, and winter ice damage. The production of the lake as a whole is the result of production from plants growing in the littoral zone, combined with production from plankton growing in the open water.

Wetlands can be part of the lentic system, as they form naturally along most lake shores, the width of the wetland and littoral zone being dependent upon the slope of the shoreline and the amount of natural change in water levels, within and among years. Often dead trees accumulate in this zone, either from windfalls on the shore or logs transported to the site during floods. This woody debris provides important habitat for fish and nesting birds, as well as protecting shorelines from erosion.

Two important subclasses of lakes are ponds, which typically are small lakes that intergrade with wetlands, and water reservoirs. Over long periods of time, lakes, or bays within them, may gradually become enriched by nutrients and slowly fill in with organic sediments, a process called succession. When humans use the watershed, the volumes of sediment entering the lake can accelerate this process. The addition of sediments and nutrients to a lake is known as eutrophication.

Ponds

Ponds are small bodies of freshwater with shallow and still water, marsh, and aquatic plants. They can be further divided into four zones: vegetation zone, open water, bottom mud and surface film. The size and depth of ponds often varies greatly with the time of year; many ponds are produced by spring flooding from rivers. Food webs are based both on free-floating algae and upon aquatic plants. There is usually a diverse array of aquatic life, with a few examples including algae, snails, fish, beetles, water bugs, frogs, turtles, otters and muskrats. Top predators may include large fish, herons, or alligators. Since fish are a major predator upon amphibian larvae, ponds that dry up each year, thereby killing resident fish, provide important refugia for amphibian breeding. Ponds that dry up completely each year are often known as vernal pools. Some ponds are produced by animal activity, including alligator holes and beaver ponds, and these add important diversity to landscapes.

Lotic

The major zones in river ecosystems are determined by the river bed's gradient or by the velocity of the current. Faster moving turbulent water typically contains greater concentrations of dissolved oxygen, which supports greater biodiversity than the slow moving water of pools. These distinctions form the basis for the division of rivers into upland and lowland rivers. The food base of streams within riparian forests is mostly derived from the trees, but wider streams and those that lack a canopy derive the majority of their food base from algae. Anadromous fish are also an important source of nutrients. Environmental threats to rivers include loss of water, dams, chemical pollution and introduced species. A dam produces negative effects that continue down the watershed. The most important negative effects are the reduction of spring flooding, which damages wetlands, and the retention of sediment, which leads to loss of deltaic wetlands.

Wetlands

Wetlands are dominated by vascular plants that have adapted to saturated soil. There are four main types of wetlands: swamp, marsh, fen and bog (both fens and bogs are types of mire). Wetlands are the most productive natural ecosystems in the world because of the proximity of water and soil. Hence they support large numbers of plant and animal species. Due to their productivity, wetlands are often converted into dry land with dykes and drains and used for agricultural purposes. The construction of dykes, and dams, has negative consequences for individual wetlands and entire watersheds. Their closeness to lakes and rivers means that they are often developed for human settlement. Once settlements are constructed and protected by dykes, the settlements then become vulnerable to land subsidence and ever increasing risk of flooding. The Louisiana coast around New Orleans is a well-known example; the Danube Delta in Europe is another.

Functions

Aquatic ecosystems perform many important environmental functions. For example, they recycle nutrients, purify water, attenuate floods, recharge ground water and provide habitats for wildlife. Aquatic ecosystems are also used for human recreation, and are very important to the tourism industry, especially in coastal regions.

The health of an aquatic ecosystem is degraded when the ecosystem's ability to absorb a stress has been exceeded. A stress on an aquatic ecosystem can be a result of physical, chemical or biological alterations of the environment. Physical alterations include changes in water temperature, water flow and light availability. Chemical alterations include changes in the loading rates of biostimulatory nutrients, oxygen consuming materials, and toxins. Biological alterations include over-harvesting of commercial species and the introduction of exotic species. Human populations can impose excessive stresses on aquatic ecosystems. There are many examples of excessive stresses with negative consequences. Consider three. The environmental history of the Great Lakes of North America illustrates this problem, particularly how multiple stresses, such as water pollution, over-harvesting and invasive species can combine. The Norfolk Broadlands in England illustrate similar decline with pollution and invasive species. Lake Pontchartrain along the Gulf of Mexico illustrates the negative effects of different stresses including levee construction, logging of swamps, invasive species and salt water intrusion.

Abiotic Characteristics

An ecosystem is composed of biotic communities that are structured by biological interactions and abiotic environmental factors. Some of the important abiotic environmental factors of aquatic ecosystems include substrate type, water depth, nutrient levels, temperature, salinity, and flow. It is often difficult to determine the relative importance of these factors without rather large experiments. There may be complicated feedback loops. For example, sediment may determine the presence of aquatic plants, but aquatic plants may also trap sediment, and add to the sediment through peat.

The amount of dissolved oxygen in a water body is frequently the key substance in determining the extent and kinds of organic life in the water body. Fish need dissolved oxygen to survive, although their tolerance to low oxygen varies among species; in extreme cases of low oxygen some fish even resort to air gulping. Plants often have to produce aerenchyma, while the shape and size of leaves may also be altered. Conversely, oxygen is fatal to many kinds of anaerobic bacteria.

Nutrient levels are important in controlling the abundance of many species of algae. The relative abundance of nitrogen and phosphorus can in effect determine which species of algae come to dominate. Algae are a very important source of food for aquatic life, but at the same time, if they become over-abundant, they can cause declines in fish when they decay. Similar over-abundance of algae in coastal environments such as the Gulf of Mexico produces, upon decay, a hypoxic region of water known as a dead zone.

The salinity of the water body is also a determining factor in the kinds of species found in the water body. Organisms in marine ecosystems tolerate salinity, while many freshwater organisms are intolerant of salt. The degree of salinity in an estuary or delta is an important control upon the type of wetland (fresh, intermediate, or brackish), and the associated animal species. Dams built upstream may reduce spring flooding, and reduce sediment accretion, and may therefore lead to saltwater intrusion in coastal wetlands.

Freshwater used for irrigation purposes often absorbs levels of salt that are harmful to freshwater organisms.

Biotic Characteristics

The biotic characteristics are mainly determined by the organisms that occur. For example, wetland plants may produce dense canopies that cover large areas of sediment or snails or geese may graze the vegetation leaving large mud flats. Aquatic environments have relatively low oxygen levels, forcing adaptation by the organisms found there. For example, many wetland plants must produce aerenchyma to carry oxygen to roots. Other biotic characteristics are more subtle and difficult to measure, such as the relative importance of competition, mutualism or predation. There are a growing number of cases where predation by coastal herbivores including snails, geese and mammals appears to be a dominant biotic factor.

Autotrophic Organisms

Autotrophic organisms are producers that generate organic compounds from inorganic material. Algae use solar energy to generate biomass from carbon dioxide and are possibly the most important autotrophic organisms in aquatic environments. The more shallow the water, the greater the biomass contribution from rooted and floating vascular plants. These two sources combine to produce the extraordinary production of estuaries and wetlands, as this autotrophic biomass is converted into fish, birds, amphibians and other aquatic species.

Chemosynthetic bacteria are found in benthic marine ecosystems. These organisms are able to feed on hydrogen sulfide in water that comes from volcanic vents. Great concentrations of animals that feed on these bacteria are found around volcanic vents. For example, there are giant tube worms (Riftia pachyptila) 1.5 m in length and clams (Calyptogena magnifica) 30 cm long.

Heterotrophic Organisms

Heterotrophic organisms consume autotrophic organisms and use the organic compounds in their bodies as energy sources and as raw materials to create their own biomass. Euryhaline organisms are salt tolerant and can survive in marine ecosystems, while stenohaline or salt intolerant species can only live in freshwater environments.

Tundra

Tundra is a major zone of treeless level or rolling ground found in cold regions, mostly north of the Arctic Circle (Arctic tundra) or above the timberline on high mountains (alpine tundra). Tundra is known for large stretches of bare ground and rock and for patchy mantles of low vegetation such as mosses, lichens, herbs, and small shrubs. This surface supports a meagre but unique variety of animals. The Finns called their treeless northern reaches the tunturi, but the concept of a vast frozen plain as a special ecological realm called tundra was developed by the Russians.

One constant factor shaping the tundra is alternate freezing and thawing of the ground. Along with the factors, this freeze-thaw cycle sets the tundra apart from two ecosystems frequently found adjacent to it the icy polar barrens on the one hand and the evergreen taiga on the other. Permafrost perennially frozen ground is a significant feature of the Arctic tundra; however, it does not typically occur in alpine regions.

Distribution

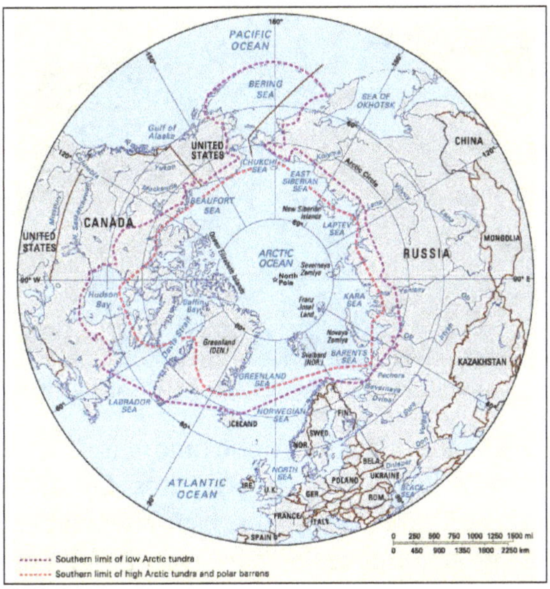

Southern limit of Arctic tundra and approximate line of demarcation between Low and High Arctic.

The global extent of the tundra biome is considerable, accounting for roughly 10 percent of Earth's surface. The southern limit of Arctic tundra follows the northern edge of the coniferous forest belt. In North America this line lies above latitude 60° N, while in Eurasia most of it occurs north of 70° N except in eastern Siberia, where it extends southward to 60° N in Kamchatka. The northward bulge of forest in Eurasia is a result of the warmer summers that occur over that large contiguous landmass.

Alpine tundra covers approximately 3 percent of Earth's land surface, and it is mostly found in the Northern Hemisphere. This habitat can be found in mountainous areas worldwide, occurring at high elevations where temperatures are too low and winds are too strong for the growth of trees. The average elevation where alpine tundra occurs is generally higher near the Equator than at the poles.

Because nearly all of Antarctica is covered with ice, it lacks a well-developed tundra, though lichens, mosses, and at least three species of flowering plants occur in more favourable habitats there.

Environmental Conditions

Climate

Tundra climates vary considerably. The most severe occur in the Arctic regions, where temperatures fluctuate from 4 °C (about 40 °F) in midsummer to −32 °C (−25 °F) during the winter months. Alpine tundra has a more moderate climate: summers are cool, with temperatures that range from 3 to 12 °C (37 to 54 °F), and winters are moderate, with temperatures that rarely fall below −18 °C (0 °F).

Aerial view of high alpine tundra near the Noatak River in the Noatak National Preserve, Alaska.

Unlike other biomes, such as the taiga, the Arctic tundra is defined more by its low summer temperatures than by its low winter temperatures. Coastal tundra ecosystems are cooler and foggier than those farther inland. Late summer and early fall are particularly cloudy seasons because large amounts of water are available for evaporation. With the first winter freeze, however, the clear skies return.

Over most of the Arctic tundra, annual precipitation, measured as liquid water, amounts to less than 38 cm (15 inches), roughly two-thirds of it falling as summer rain. The remainder falls in expanded form as snow, which can reach total accumulations of 64 cm

(25 inches) to (rarely) more than 191 cm (75 inches). Annual precipitation has a wide range in alpine tundra, but it is generally higher in Arctic tundra. For example, annual precipitation may be as much as 64 cm (25 inches) at higher elevations in the Rocky Mountains of Colorado but may be less than 7.6 cm (3 inches) in the northwestern Himalayas. Blinding snowstorms, or whiteouts, obscure the landscape during the winter months, and summer rains can be heavy. The stratification of the soil and the inclination of the alpine slopes allow for good drainage, however.

Although winds are not as strong in the Arctic as in alpine tundras, their influence on snowdrift patterns and whiteouts is an important climatic factor. Blizzard conditions developing in either location may reduce visibility to roughly 9 metres (about 30 feet) and cause snow crystals to penetrate tiny openings in clothing and buildings. Winds in the alpine tundras are often quite strong; they may average 8 to 16 km (5 to 10 miles) per hour only 60 cm (about 24 inches) above ground level, and they quite frequently reach 120 to 200 km (about 75 to 125 miles) per hour in high reaches of the Rocky Mountains and the Alps.

Soils

As noted above, permafrost is an ever-present feature of the Arctic tundra. The southern limit of continuous permafrost occurs within the northern forest belt of North America and Eurasia, and it can be correlated with average annual air temperatures of −7 °C (20 °F). South of this zone, permafrost exists in patches. Over much of the Arctic, permafrost extends to depths of 350 to 650 metres (1,150 to 2,100 feet). In unglaciated areas of Siberia, however, permafrost may reach 1,450 metres (4,760 feet).

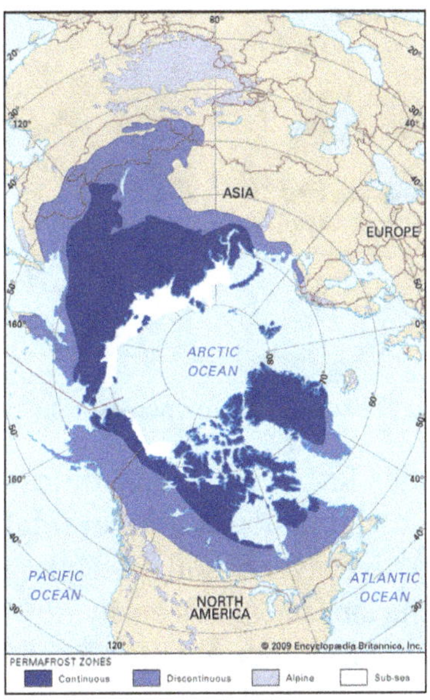

Distribution of permafrost in the Northern Hemisphere.

The presence of permafrost retards the downward movement of water though the soil, and lowlands of the Arctic tundra become saturated and boggy during the summer thaw. During the winter, water in the soil can freeze into a lens of ice that causes the ground above it to form into a hilly structure called a pingo. Alpine tundra is generally drier, even though the amount of precipitation, especially as snow, is higher than in Arctic tundra. In alpine tundra the lack of a continuous permafrost layer and the steep topography result in rapid drainage, except in certain alpine meadows where topography flattens out.

Pingo. Pingos near Tuktoyaktuk, Northwest Territories, Canada.

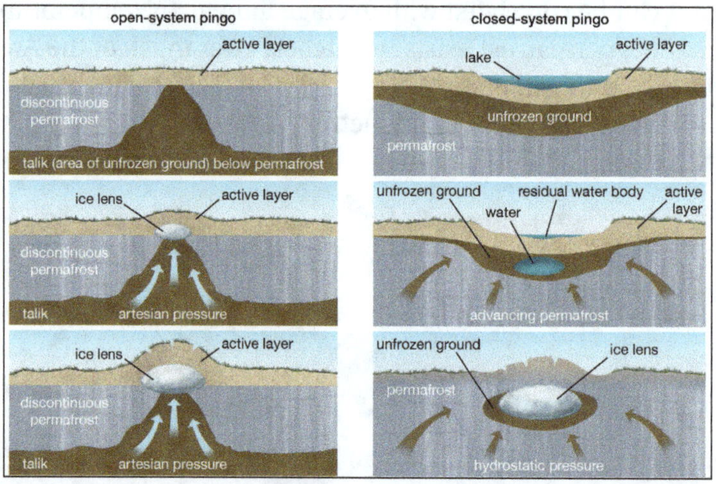

Pingo. Diagram explaining the formation of open-system and closed-system pingos.

Patterned ground, a conspicuous feature of most tundras, results from the differential movement of soil, stone, and rock on slopes and level land, plus the downward creep (solifluction) of the overlying active layer of soil. (Because permafrost is impermeable to water, waterlogged soil near the surface slides easily down a slope.) These phenomena are a result of the freeze-thaw cycle common to the tundra and are especially common in spring and fall. In alpine regions, surface features such as rock rings, stripes, and polygons are seen, usually measuring 15 to 30 cm (6 to 12 inches) across. Where there is adequate moisture for soil lubrication, solifluction terraces and lobes are common. In the Arctic tundra, solifluction is often cited as the reason why rock slabs may be found standing on end.

Thawed surface of the permafrost on the tundra in summer, Taymyr Peninsula, Siberia.

Tundra soils are usually classified as Gelisols or Cryosols, depending on the soil classification system used. Both are easily eroded soil types characterized by the presence of permafrost and showing an active surface layer shaped by the alternating freezing and thawing that comes with seasonal variations in temperature. In the higher latitudes of the Arctic, the summer thaw penetrates to a depth of 15 to 30 cm (6 to 12 inches). In lower latitudes characterized by full plant cover and well-drained soils, the thaw penetrates from 0.5 to 3 metres (1.5 to 10 feet). Most biological activity, in terms of root growth, animal burrowing, and decomposition of organic matter, is limited to the active layer. Although the permafrost layer exists only in Arctic tundra soils, the freeze-thaw layer occurs in soils of both Arctic and alpine tundra.

Gelisol soil profile showing a year-round frozen subsurface layer (permafrost) below a dark surface horizon rich in organic matter.

Cryosol soil profile from Canada showing patterned deformations caused by cycles of freeze and thaw.

The Biota and its Adaptations

In Arctic and alpine tundras, the number of species of plants and animals is usually small when compared with other regions, yet the number of individuals per species is often high. Food and feeder relationships are simple, and they are more subject to upset if a critical species disappears or decreases in number. Many tundra species cannot be found elsewhere, and thus the biome is an important contributor to global biodiversity despite its low species number.

Cotton grass (Eriophorum) and mosses in the Arctic tundra, northern Yukon, Canada.

Musk oxen grazing in Bering Land Bridge National Preserve, western Alaska, U.S.

The tundra also hosts abundant bacteria and fungi, which are essential to proper ecosystem functioning in the biome. Several studies using DNA sequencing and analysis have discovered many novel microbial groups in tundra soils. These microbial communities are active under the snow, and their composition changes dramatically from winter and spring to summer in response to changes in soil temperature, moisture, carbon availability, and the nature of carbon-containing substrates (the surfaces upon which microbes live). Microbes and fungi play a key role in biogeochemical processes, such as nutrient regeneration and the carbon cycle.

Plant Life

The vegetation of many alpine tundras and over most of the Arctic tundra tends to be greenish brown in colour. While plants do not remain in flower for more than a few days or weeks in these environments, the blossoms are generally large in relation to the size of the plant and are rather colourful, especially in alpine habitats. The foggy tundras found along coastal areas produce matted and grassy landscapes. Algae and fungi are found along rocky cliffs, and rosette plants grow in rock cornices and shallow gravel beds. In the drier inland tundras, spongy turf and lichen heaths develop.

Wildflowers blooming on the tundra in Arctic National Wildlife Refuge, northeastern Alaska, U.S.

Across the southerly Arctic tundra, which is marked by vast areas of low relief, boggy peat soils with an abundance of lakes and meandering rivers prevail. These coastal plain areas are dominated by sedges and cotton grass, and mosses including Sphagnum are common. On slightly elevated sites, often only 15 to 60 cm (6 to 24 inches) above the wet peaty soils, low willows (Salix), grasses, and rushes occur. Taller willows, grasses, and plants in the sunflower and legume families (Asteraceae and Leguminosae, respectively) are common on the sands and gravels of riverbanks, but vegetation is quite sparse on higher lands, foothills, and Arctic mountains.

In Arctic and alpine tundra ecosystems, the plant communities are influenced by soil drainage, snow cover and time of melt, and localized microclimates that differ from one another in temperature, wind, soil moisture, and nutrients. On gentle slopes where soil has developed, extensive meadows occur. On windswept ridges, cushion plants dominate. On rocky slopes and peaks, plants are found in scattered patches where there is a bit of soil and some snow cover in winter. In higher mountains, having much snow, ice, and exposed rock, lichens and mosses manage to grow on rocks. Vascular plants usually end at or just below the line of permanent snow.

The transition from mountain forest to the shrub- and herb-dominated alpine tundra at higher elevations is very similar to the transition from the coniferous forest belt to the Arctic tundra at higher latitudes. The alpine transition, however, occurs over only 100 metres (330 feet) or so of vertical rise. Timberline trees are mostly spruce (Picea), fir (Abies), and pine (Pinus), with very few deciduous tree species. Willow clumps less than 60 cm (about 24 inches) tall are common in the krummholz (a transitional zone of scattered clusters of stunted trees) and beyond, where snowdrifts are extensive. Willows are also common along streams, in the lee of rocks, and in basins or on the lee side of ridges where winter snow is deeper.

With plant growth and many aspects of animal activity confined to two to four months of the year, when temperatures are above freezing, evolution has favoured a rapid completion of life cycles. Tundra organisms are opportunistic. Many species of plants are perennials that flower within a few days after the snow begins to melt, and some produce ripe seed within four to six weeks. Very few species are annuals. Plants 2.5 to 7.5 cm (1 to 3 inches) tall typically flower first, because they are in the warmer air layers near the soil surface. During cloudy periods, in shade, and at night, flower temperature is very similar to that of the surrounding air. In sunlight, however, flowers may be about 2–10 °C (4–18 °F) warmer than the air around them. Some plants that freeze while in flower when sudden storms hit continue to develop and produce seed upon thawing. Many plants set few seeds and depend mostly upon runners or underground stems for increasing their numbers, such as a number of Arctic species in the heath family (Ericaceae). A few species produce bulblets that develop roots and shoots on the parent plant before they drop to the ground. The hairy flower stalks of cottongrass (Eriophorum), lousewort (Pedicularis), and willows retain warm air, raising the temperature near the stalks by 3–9 °C (5–15 °F); this ability is an important adaptation for flowering in areas where air temperatures may approach the freezing point.

Animal Life

Organisms of the northern alpine tundra probably evolved before those of the Arctic tundra, appearing first in the Mongolo-Tibetan Plateau. Few alpine animals, however, contributed directly to the evolution of Arctic tundra species, because physical barriers prevented the migration of species and because alpine and Arctic animals were specialized to their particular environments. However, alpine plants and some animals migrated east and west through mountain ranges to Europe and North America.

Tetraonidae.
Ptarmigan (Lagopus muta).

Meadow vole.

The meadow vole (Microtus pennsylvanicus) is one of the most common mammals in North America. Its geographic range spans most of Canada and the conterminous United States as well as most of Alaska.

Lowland tundra animals appear to have evolved in central Eurasia when tundra replaced the cold temperate steppe. These animals migrated west to Europe about one million years ago, during the middle of the Pleistocene Epoch, and later migrated east to North America across the Bering Land Bridge. As a product of this migration, many of the common Arctic animals are circumpolar. Such animals include the polar bear (Ursus maritimus), Arctic fox (Alopex lagopus), Arctic wolf (Canis lupus), Arctic hare (Lepus arcticus), Arctic weasel (Mustela), snowy owl (Nyctea scandiaca), several species of lemming (a group of rodents that make up part of the subfamily Arvicolinae) and ptarmigan (Lagopus), and a number of species of waterfowl.

Warm interglacial periods (that is, the intervals between ice ages) eliminated many tundra species and were probably detrimental to the reappearance of truly alpine animals. Thus, the Alps, Rocky Mountains, and other ranges today have few fauna that are strictly limited to the alpine tundra. Larger animals for example, the chamois spend the more favourable seasons above the timberline but then descend the slopes to graze in the lowlands in winter.

Mammals

Arctic Mammals

Small mammals of the Arctic tundra have high reproductive rates. Most notable in this regard are the lemmings, which reach a population peak every three to five years in some regions. Lemmings remain active all winter, living under the snow where they feed upon the roots of grasses and sedges; they may even reproduce under the thin yet insulative snow layer. When the lemming population increases, many plants are consumed, and there is a large accumulation of feces. The buildup of manure around animal burrows in turn stimulates plant growth by adding nitrogen and other nutrients to the soil.

With environmental extremes as pronounced as they are in the Arctic and with the number of species so limited, there are often considerable oscillations in animal populations. Lemmings are the most prominent example, but the populations of animals that prey on lemmings such as jaegers (Stercorarius), snowy owls, and foxes also rise and fall, closely following the rise and fall of their prey. Snowy owls simply migrate to the coniferous forest belt when lemmings are scarce, but fox populations drop significantly. On the other hand, when peaks in the lemming population occur, vegetation becomes scarce, and large numbers of lemmings move into less densely populated areas.

The characteristic large herbivores of the Arctic tundra are the reindeer (Rangifer tarandus) of Eurasia and North America (where they are known as caribou) and the musk ox (Ovibos moschatus) of Greenland and some Canadian Arctic islands. These animals are rather large, considering the severe environments in which they live. Greater body size confers an adaptive advantage: there is less surface area relative to volume and, therefore, less opportunity for heat to dissipate to the outside. Musk oxen are especially well equipped for surviving in cold climates because of their exceptionally thick coats. Reindeer possess sharp hooves and antlers, which enable them to cut through snow to forage on lichens and flowering plants.

Some predatory mammals follow the seasonal patterns of their prey, whereas others obtain food near their dens. Polar bears are as much a part of the marine environment as they are of the tundra. In winter they spend much of their time on ice floes hunting seals and other animals that congregate in those locations to breed. Brown bears (Ursus arctos) capture fish and small mammals, and foxes and wolves often follow bears to feed on leftovers. In summer, foxes and wolves are found mostly on land, where they raise their young and feed on birds and small mammals.

Many tundra animals sport white coats in winter as part of distinct summer-winter phase shifts in coloration; among them are the foxes, Arctic hares, and ptarmigans. This camouflage helps both predator and prey: predators can steal up without detection, and prey can hide easily in the snow. The white coat of polar bears is caused by the animal's specialized transparent hairs that reflect visible light.

Alpine Mammals

Alpine tundras are similarly limited in the number of animal species and diversity. In this case, however, the limited fauna may be partly due to the fact that many mammals and other animals found in alpine tundras are not adapted for year-round alpine life. Instead, they follow a vertical migration, descending into forested environments in winter and returning to the heights in summer. Migratory animals include mountain sheep (Ovis canadensis), ibex (Capra), chamois (Rupicapra), several wildcats, and many birds. Mountain goats (Oreamnos americanus) spend noticeably more time at higher elevations in winter.

Mount Evans; Colorado.
Mountain goats along the side of the road up Mount Evans, west of Denver, Colorado, U.S.

In contrast with Arctic tundra mammals, some alpine mammals hibernate during the winter. Marmots (Marmota), ground squirrels, and jumping mice and other rodents of family Zapodidae consume large amounts of vegetation in summer and early fall before hibernation begins. Other small mammals, such as pikas (Ochotona) and voles (a group of rodents similar to lemmings that make up part of the subfamily Arvicolinae), cache hay in the fall for winter feeding. Rabbits and others forage as they can in winter, and foxes range over large areas of alpine habitats.

Birds

Most tundra birds are migratory, staying long enough to nest and molt. One exception is the ptarmigan, which feeds upon willow buds and other exposed plant parts in winter and upon leaves, buds, and flowers in summer. Ptarmigan have heavily feathered feet which provide some insulation against the winter snow and ice. Several migratory birds feed upon seeds and fruits until insects and spiders emerge and become available in summer. Some migratory birds, such as the snow goose (Chen caerulescens), alter the landscape. Snow geese often denude areas of cotton grass, leaving behind mostly mosses, which increases the flux of solar energy into soils. In this way, they indirectly promote deep thaws that may result in soil creep on slopes. Important birds of prey are the jaegers, which are summer visitors, and the snowy owls, which are year-round residents, though the latter move southward into the forest in winters when food supply is scarce. The several species of jaegers and the owls feed upon small birds and insects, although lemmings are the most important item of their diet.

Snowy owl (Bubo scandiacus) in flight.

Snow geese (Chen caerulescens) flying in V-formation.

Insects

Although the number of insect species in the Arctic is small compared with that of temperate regions, those that are present are quite successful. Arctic insects resist freezing winter temperatures. Some groups of species, such as tundra-adapted mosquitoes, possess high concentrations of glycerol, which acts as an "antifreeze" to lower the temperature at which freezing occurs. Many tundra insects and spiders are dark in colour, which absorbs more sunlight and allows these animals to maintain higher body temperatures. Some of the tundra species of black flies and mosquitoes do not require a blood meal before depositing their eggs, in contrast to their temperate-region counterparts.

Use of Sunlight and Carbon Dioxide

The flora and fauna of Arctic tundras and alpine tundras are affected by differences in day length and in the concentration of carbon dioxide (CO_2) in the atmosphere. Plants and animals of alpine tundras are subjected to the same day-night regime as are other organisms at lower elevations in tropical or temperate regions, and numerous activities of these organisms are controlled by the length of the night. Over most of the Arctic, on the other hand, light prevails continuously for one to four months, and biological rhythms are induced by variables other than a daily dark period. Many Arctic tundra plants, for instance, flower abundantly only when exposed to continuous or nearly continuous sunlight. In insects, the rhythms of feeding, flight, and swarming, which are normally controlled by light-dark cycles, respond rather to prevailing temperature or sunlight. Birds and large mammals of the Arctic tundra appear to observe a "quiet period" in early morning, though this period is not as pronounced as it is in the animals that inhabit alpine tundras in temperate regions. CO_2 levels are lower in alpine tundras because of thinner air at higher altitudes, and alpine plants are more efficient in utilizing these lower levels of CO_2 in photosynthesis than are their Arctic counterparts.

Biological Productivity

An important measure of natural ecosystems is the biological production of its plants and animals—that is, the total amount of biomass produced by living organisms within a given area in a specific period of time. In polar regions the greatest biological production occurs in marine waters rather than on land, and production is actually higher in the Antarctic than it is in the Arctic Ocean. Production studies of Arctic tundra lakes indicate that there are many species of algae and aquatic mosses living in the higher latitudes of the Arctic.

Rates of annual plant productivity in the tundra vary from less than 10 grams per square metre, or 0.003 ounce per square foot (dry weight) in polar desert (or frigid desert) environments to 400 grams per square metre (about 1.3 ounces per square foot) in some alpine sedge meadows. Arctic shrub tundra has a range of 100–300 grams per square metre (0.3–0.98 ounces per square foot). For comparison, tropical rainforests produce 1,000–4,000 grams per square metre (3.3–13.1 pounds per square foot). While other environments benefit from longer growing seasons, the growing season in tundra habitats is extremely short, often less than two months per year. On the other hand, the short growing season and the low solar angle in the Arctic are balanced by long summer days, which allow plants to carry out photosynthesis for a full 24 hours per day.

These data do not tell the entire story, however, because the amount of available plant material (standing crop) is often 10 to 25 times greater below ground than it is above ground. This indicates that the soil environment may be more favourable for the growth of plants than the air. In many tundras, the harvest by plant-eating animals accounts for no more than 0.1 to 2.0 percent of the live plants per year. This means that most of the plants produced decay and are decomposed by microorganisms. With alpine and Arctic vegetation so sparse, it is no accident that humans living in these environments have harvested herbivorous animals, including reindeer, ducks, and geese, rather than native or cultivated plants.

Effects of Human Activities and Climate Change

Earth's tundra regions are harsh and remote, so fewer humans have settled there than in other environments. However, humans have a long history in the tundra. For example, the first people who went to North America from Asia more than 20,000 years ago traveled through vast tundra settings on both continents. Since then human activity in tundra ecosystems has increased, mainly through the procurement of food and building materials. Humans have changed the landscape through the construction of residences and other structures, as well as through the development of ski resorts, mines, and roads. Hunting, oil drilling, and other activities have polluted the environment and have threatened wildlife in tundra ecosystems. Environmental scientists are concerned that the continued expansion of these activities along with the release of air pollutants, some of which deplete the ozone layer, and greenhouse gases, which hasten climate change has begun to affect the very integrity and sustainability of Arctic and alpine tundra ecosystems. For example, the increased occurrence of tundra fires would decrease the coverage of lichens, which could, in turn, potentially reduce caribou habitats and subsistence resources for other Arctic species.

The effects of climate change on tundra regions have received extensive attention from scientists as well as policy makers and the public. This attention partly stems from the tundra's high sensitivity to the general trend of global warming. While the average global surface-air temperature has risen by approximately 0.9 °C (about 1.5 °F) since 1900, average surface air temperatures in the Arctic have risen by 3.5 °C (5.3 °F) over the same period. Many parts of the region have experienced several consecutive years of record-breaking winter warmth since the late 20th century. In some locations, this record-breaking winter warmth has been unprecedented; three-month winter mean temperatures in Norway's Svalbard archipelago in 2016 were 8–11 °C (14.4–19.8 °F) higher than the 1961–90 average. Most climatologists agree that this warming trend will continue, and some models predict that high-latitude land areas will be 7–8 °C (12.6–14.4 °F) warmer by the end of the 21st century than they were in the 1950s.

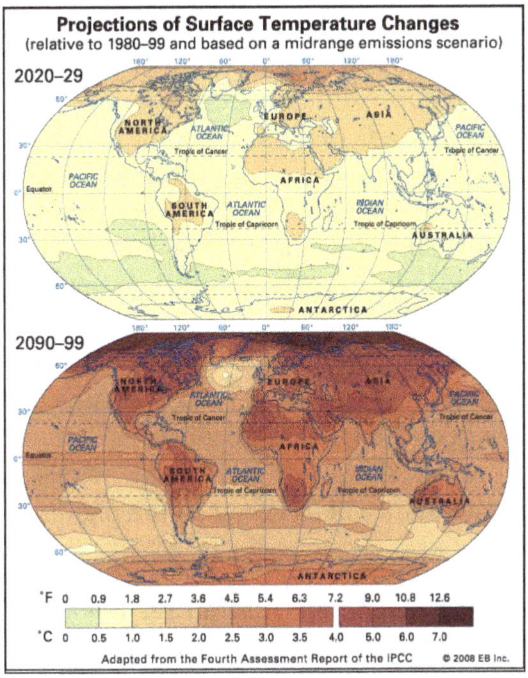

Global warming has already produced detectable changes in Arctic and alpine tundra ecosystems. These ecosystems are being invaded by tree species migrating northward from the forest belt, and coastal areas are being affected by rising sea levels. Both phenomena are reducing the geographic extent of the Arctic tundra. Other changes occurring in both Arctic and alpine tundras include increased shrub density, an earlier spring thaw and a later autumn freeze, diminished habitats for native animals, and an accelerated decomposition of organic matter in the soil. These processes can actually contribute to greater warming in the tundra than in other regions. For example, climatologists point out that the darker surfaces of green coniferous trees and ice-free zones reduce the albedo (surface reflectance) of Earth's surface and absorb more solar radiation than do lighter-coloured snow and ice, thus increasing the rate of warming.

One of the most striking ongoing changes in the Arctic is the rapid melting of sea ice. Some climate models predict that, sometime during the first half of the 21st century, summer sea ice will vanish from the Arctic Ocean. An absence of summer ice would amplify the existing warming trend in Arctic tundra regions as well as in regions beyond the tundra, because sea ice reflects sunlight much more readily than the open ocean and, thus, has a cooling effect on the atmosphere. In addition, research indicates that the retreat of sea ice would enhance the productivity of tundra vegetation, and the resulting buildup of plant biomass might lead to more extreme events such as large tundra fires. Finally, an ice-free Arctic Ocean would improve access to high northern latitudes for recreational and industrial activities; this would likely place additional stress on tundra plants and animals as well as compromise the resilience of the tundra ecosystem itself. In alpine tundras too, climate warming could encourage more human activity and increase damage to plant and animal populations there.

The fate of permafrost in a warmer world is a particularly important issue. Together, tundra and taiga account for approximately one-third of global carbon storage in soil, and a large portion of this carbon is tied up in permafrost in the form of dead organic matter. Some of this organic matter has been preserved for many thousands of years, not because it is inherently difficult to break down but because the land has remained frozen. Thawing of the permafrost would expose the organic material to microbial decomposition, which would release carbon into the atmosphere in the form of CO_2 and methane (CH_4). Rates of microbial decomposition are much lower under anaerobic conditions, which release CH_4, than under aerobic conditions, which produce CO_2; however, CH_4 has roughly 25 times the greenhouse warming potential of CO_2. The Arctic has been a net sink (or repository) of atmospheric CO_2 since the end of the last ice age. At the same time, however, the region has been a net source of atmospheric CH_4, primarily because of the abundance of wetlands in the region.

Numerous other factors affect the exchange of carbon-containing compounds between the tundra and the atmosphere. Tundra fires release CO_2 to the atmosphere, and there is evidence that climate warming over the past several decades has increased the frequency and severity of tundra burning in the Arctic. In contrast, greater plant productivity resulting from a longer, warmer growing season could compensate for some of the carbon emissions from permafrost melting and tundra fires. Indeed, ecologists and climate scientists note that there is a great deal of uncertainty about the future of the carbon cycle in the Arctic during the 21st century. They worry, however, that a net transfer of greenhouse gases from tundra ecosystems to the atmosphere has the potential to exacerbate changes in Earth's climate through a positive feedback loop, in which small increases in air temperature at the surface set off a chain of events that leads to further warming.

Burning tundra. A fire burning across a landscape of forest and tundra in northwestern Alaska.

Ecoregion

Ecoregions or bioregions are geographical units with characteristic flora, fauna and ecosystems. They are a division of the larger "ecozones" or biogeographic regions. Political divisions of counties, states and countries do not respect ecological processes, so it is important to develop public policies that include appropriate spatial scales and that consider environmental dynamics. The ecoregional approach is important from several points of view.

Many species require large areas to maintain viable populations. For example, raptors and large predators have areas of activity of more than 20 km^2. To keep populations of hundreds of individuals of these species, we need large areas.

Ecosystem processes, such as the cycles of water and nutrients, sometimes operate at large spatial and temporal scales, e.g. the composition of coastal lagoons is influenced by the ocean and by the supply of currents of fresh water that come from great distances.

Most human communities perform their economic and social activities within a regional environmental setting. A healthy environment provides goods and services of high quality to the populations within. By contrast, large cities go beyond their local or regional scope in order to satisfy the needs of their residents: water, food, materials and energy must be imported from distant places.

Phytochorion

A phytochorion, is a geographic area with a relatively uniform composition of plant species. Adjacent phytochoria do not usually have a sharp boundary, but rather a soft one, a transitional area in which many species from both regions overlap. The region of overlap is called a vegetation tension zone.

In traditional schemes, areas in phytogeography are classified hierarchically, according to the presence of endemic families, genera or species, e.g., in floral (or floristic, phytogeographic) zones and regions, or also in kingdoms, regions and provinces, sometimes including the categories empire and domain.

Systems used to classify vegetation can be divided in two major groups: those that use physiognomic-environmental parameters and characteristics and those that are based on floristic (i.e. shared genera and species) relationships. Phytochoria are defined by their plant taxonomic composition, while other schemes of regionalization (e.g., vegetation type, physiognomy, plant formations, biomes) may variably take in account, the apparent characteristics of a community (the dominant life-form), environment characteristics, the fauna associated, anthropic factors or political-conservationist issues.

Several systems of classifying geographic areas where plants grow have been devised. Most systems are organized hierarchically, with the largest units subdivided into smaller geographic areas, which are made up of smaller floristic communities, and so on. Phytochoria are defined as areas possessing a large number of endemic taxa. Floristic kingdoms are characterized by a high degree of family endemism, floristic regions by a high degree of generic endemism, and floristic provinces by a high degree of species endemism. Systems of phytochoria have both significant similarities and differences with zoogeographic provinces, which follow the composition of mammal families, and with biogeographical provinces or terrestrial ecoregions, which take into account both plant and animal species.

The term "phytochorion" is especially associated with the classifications according to the methodology of Josias Braun-Blanquet, which is tied to the presence or absence of particular species, mainly in Africa.

Taxonomic databases tend to be organized in ways which approximate floristic provinces, but which are more closely aligned to political boundaries, for example according to the World Geographical Scheme for Recording Plant Distributions.

Barren Vegetation

Barren vegetation in Luzerne County, Pennsylvania, USA.

Barren vegetation describes an area of land where plant growth may be sparse, stunted, and/or contain limited biodiversity. Environmental conditions such as toxic or infertile soil, high winds, coastal salt-spray and climatic conditions are often key factors in poor plant growth and development. Barren vegetation can be categorized depending on the climate, geology and the geographic location of a specific area. Pine barrens, coastal barrens and serpentine barrens are some of the more distinct ecoregions for barren vegetation and are the most commonly researched by scientists. Often referred to as "heathlands", barrens can be excellent environments for unique biological diversity and taxonomic compositions.

Serpentine Barrens

Biological Diversity

Serpentine barren habitats include grasslands, chaparral, and woodlands as well as some areas that are very sparsely vegetated. Areas of sparse vegetation are often characterized by annual and perennial herbaceous plant species. The flora of the serpentines is recognized globally for its high level of biological diversity which includes over 1600 taxa of plants occurring in serpentine areas of the eastern U.S., with as many as 2000 taxa considered to be endemic to serpentine rich soils.

Geology

Serpentine barrens are distinct due to the serpentine-rich soil produced by the hydration weathering and metamorphic transformation of ultramafic igneous bedrock. Serpentine barrens are often characterized as high-stress environments with low water and nutrient availability. These areas are often depleted in basic nutrients such as nitrogen and phosphorus. The soil is often shallow and can be toxic due to high heavy metal concentrations such as nickel, cobalt and chromium. As a result of the harsh conditions and unique edaphic properties presented by serpentine barrens these environments support stress-tolerant plant communities characterized by distinct and locally defined plant species.

Pine Barrens

Biological Diversity

The ecoregion known as the Pine Barrens are found to spread across much of the northeastern United States, primarily in the state of New Jersey. The Pine Barrens comprise 550,000 hectares of a heavily forested area of coastal plain and are home to at least 850 species of plant life, including many which are endangered or threatened.

Geology

The Pine Barrens are primarily formed on unconsolidated, acidic, medium-to-coarse grained sands and gravel. The mature soils are considered to be true Podzols and are siliceous and highly permeable. The low moisture holding capacity and nutrient status of the soil create low vegetation growth rates throughout much of the Pine Barrens.

Avalon Peninsula Newfoundland Canada.

Coastal Barrens

Coastal Barrens are characterized by short vegetation, sparse tree cover, exposed bedrock, and bog pockets. Often, coastal barrens exhibit stressful climatic conditions and are subjected to consistent windy conditions and salt-spray. Along the Atlantic coast of Nova Scotia and northeastern United States, there are patches of unforested coastal barrens spread throughout areas that contain exposed bedrock and/or little soil cover within a forested landscape. More extensive barrens can be found in much of Newfoundland and Labrador and further north in mainland Canada.

Biological Diversity

Coastal Barrens typically host low growing shrub communities with sparse tree cover and are often dominated by ericaceous species such as the black huckleberry (Gaylussacia baccata) and low bush blueberry (Vaccinium angustifolium). The coastal barrens of Atlantic Canada host a variety of taxonomic species such as macro lichens, mosses, and vascular plants. Studies have recorded 173 different species in various coastal barren regions of the province of Nova Scotia. This number included 105 vascular plants, 41 macro lichens, and 27 moss species with six provincially rare vascular species that were found predominantly in nearshore areas that contained high levels of substrate salt and nutrients, variable substrate depth, and short vegetation.

Climate

Barren lands are located in areas with a climate that is very damp and humid. For instance, the Buck Creek barrens in North Carolina, in the last ten years the zone receives around 1770 mm of precipitation. The Nottingham Serpentine Barren are very humid and has an average temperature of 11 degrees Celsius. Here, the average precipitation averages at 1200mm and is spread out evenly throughout the year. Another region of barren vegetation is located in the Appalachian Mountains. In the low elevation of the northern part of this mountain chain, the annual precipitation is a little lower than on the southern Appalachians high peaks. In this location, precipitation falls mostly as rain rather than snow; and also precipitation occurs mostly in the summer.

Anthropogenic Relationships

Anthropogenic interactions have been used over the years to help change and drive vegetation in the eastern US. Meaning that the actions of human-beings will play a role in what type of vegetation will grow in some locations. This is including things like fires and fire suppression, grazing, logging and agriculture clearing. Research has been done and anecdotal evidence has been shown to suggest vegetation structures and composition in the eastern serpentine barrens may have also been influenced by local disturbance regimes associated with these events as well as mining.

Savannahs and barrens are ecosystems that are rare in North America. This is due in part to human impacts, such as agriculture, urbanization, and altering the natural fire regimes. Over the past 50 years, the area of savannah-like openings and pine woodland has been continuously reduced of the years, a tendency opposite to that of hardwood forests. These changes in vegetation structures along with the composition are caused by, in part to anthropogenic changes in the fire regime. Following burning of vegetation there is a release of inorganic nutrients into the ecosystem caused by the combustion of the plants biomass and therefore, releasing the nutrients. This release of nutrients, after the occurrence is thought to be a reason for an increase in plant productivity.

Global Distribution and Geography

Regions on the earth's surface where soils are dominating the ecosystems with little to no plant cover are often referred to as "Barren". These places are areas like deserts, Polar Regions, areas of high elevation, and zones of glacier retreat. For barren zones that are situated in mountain ranges, they are often called the "Subnival Zone", and are found at elevations between the upper limit of the vegetation zone and the lower limit of the ice covered zone. Subnival zones in places like the Rockies, Andes and Himalayas have increased greatly in the past few years due to the retreat of high elevation glaciers and the ice caps. One area for study is The Nottingham Serpentine Barrens, which covers 200 ha in southern Chester Country, Pennsylvania on the Pennsylvania-Maryland border. The typical serpentine barren is either a prairie or savannah grassland. The soils here in this location are a section of the Neshaminy-Chrome-Conowingo association. These soils are deep and are derived from the serpentine bedrock. This series of soils are well drained and also moderately sloping. With this, these specific locations have been under heavy erosion forces and have a depth to its parent bedrock within a distance of 15–75 cm. Here, there is also low permeability which makes it difficult for plants to have availability to water and therefore hard to collect moisture.

Mean elevation and elevation range limits both vegetation zones and individual species should be define with increasing latitude. For example, in the southern Appalachians, high-elevation outcrops, composition gradients are a function of elevation, potential solar radiation, geographic gradient that corresponds to broad geological differences (mafic rocks to the northwest vs. felsic rocks in the southwest direction), and surficial geomorphology (bedrock surfaces that are less fractured in the southeast).

Steppe

In physical geography, a steppe is an ecoregion, in the montane grasslands and shrublands and temperate grasslands, savannas and shrublands biomes, characterized by grassland plains without trees apart from those near rivers and lakes. The prairie of North America (especially the shortgrass and mixed prairie) is an example of a steppe, though it is not usually called such. A steppe may be semi-arid or covered with grass or shrubs or both, depending on the season and latitude. The term is also used to denote the climate encountered in regions too dry to support a forest but not dry enough to be a desert. The soil is typically of chernozem type.

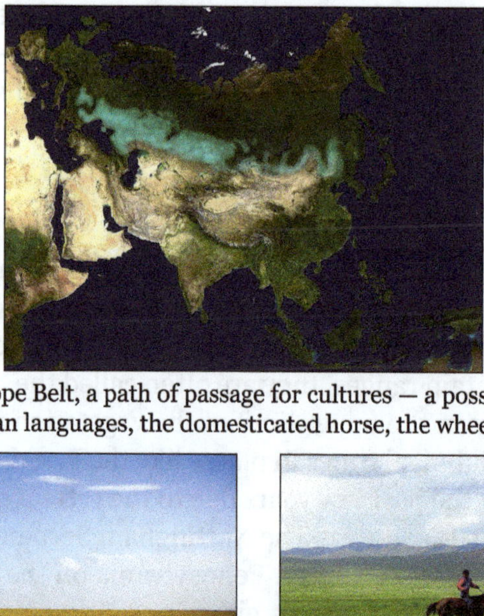

The Eurasian Steppe Belt, a path of passage for cultures — a possible origin for the Indo-European languages, the domesticated horse, the wheel and chariot.

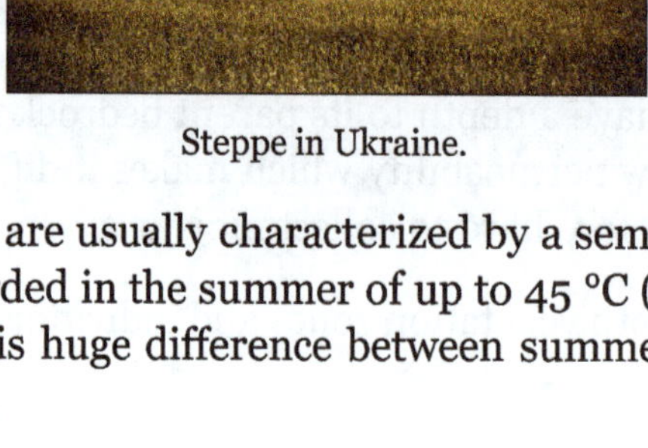

Steppe in Ukraine.

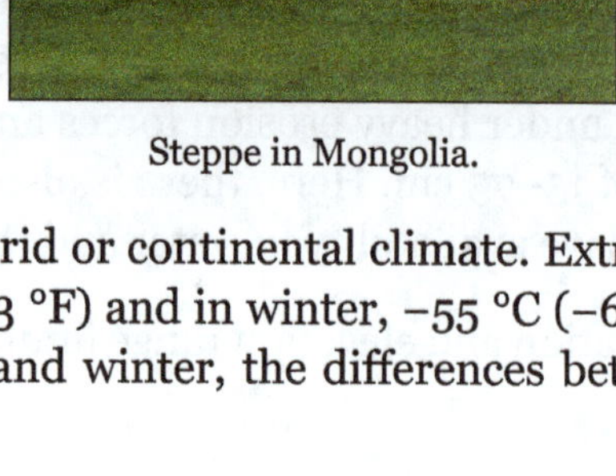

Steppe in Mongolia.

Steppes are usually characterized by a semi-arid or continental climate. Extremes can be recorded in the summer of up to 45 °C (113 °F) and in winter, −55 °C (−67 °F). Besides this huge difference between summer and winter, the differences between day and night are also very great. In both the highlands of Mongolia and northern Nevada, 30 °C (86 °F) can be reached during the day with sub-zero °C (sub 32 °F) readings at night.

Steppe in Russia.

The mid-latitude steppes can be summarized by hot summers and cold winters, averaging 250–510 mm (10–20 in) of precipitation per year. Precipitation level alone is not what defines a steppe climate; potential evapotranspiration must also be taken into account.

Two Types

Southern Siberian steppe: windbreaker trees in the wintertime.

Steppe can be classified by climate:

- Temperate steppe: The "true" steppe, found in continental areas of the world; they can be further subdivided, as in the Rocky Mountains Steppes.
- Subtropical steppe: A similar association of plants that can be found in the driest areas with a Mediterranean-like climate; it usually has a short wet period.

It can also be classified by vegetation type, e.g. shrub-steppe and alpine-steppe.

The Eurasian Grass-Steppe of the temperate grasslands, savannas, and shrublands had a role in the spread of the horse, the wheel, and the Indo-European languages. The Indo-European expansion and diverse invasions of horse archer civilizations of the steppe eventually led to the rise of Mycenaean Greece by amalgamation of Indo-Europeans with the autochthonous pre-Greek population and also its destruction during the Dorian invasion in the Late Bronze Age collapse, followed by the demise of the Achaeans, the spread of the Sea Peoples, and eventually the rise of Archaic and ultimately Classical Greece.

Locations

Cold Patagonian steppe near Fitz Roy, Argentina.

Cold Steppe

The world's largest steppe region, often referred to as "the Great Steppe", is found in Eastern Europe and Central Asia, and neighbouring countries stretching from Ukraine in the west through Russia, Kazakhstan, China, Turkmenistan and Uzbekistan to the Altai, Koppet Dag and Tian Shan ranges.

The inner parts of Anatolia in Turkey, Central Anatolia and East Anatolia in particular and also some parts of Southeast Anatolia, as well as much of Armenia and Iran are largely dominated by cold steppe.

The Pannonian Plain is another steppe region in eastern Europe, primarily Hungary.

Prairie in Alberta, Canada.

Another large steppe area (prairie) is located in the central United States, western Canada and northern part of Mexico. The shortgrass prairie steppe is the westernmost part of the Great Plains region. The Channeled Scablands in Southern British Columbia and Washington State is an example of a steppe region in North America outside of the Great Plains.

In South America, cold steppe can be found in Patagonia and much of the high elevation regions east of the southern Andes.

Relatively small steppe areas can be found in the interior of the South Island of New Zealand.

Sagebrush steppe in northeastern Nevada along US 93.

Subtropical Steppe

In Europe, some Mediterranean areas have a steppe-like vegetation, such as central Sicily in Italy, southern Portugal, parts of Greece in the southern Athens area, and central-eastern Spain, especially the southeastern coast (around Murcia), and places cut off from adequate moisture due to rain shadow effects such as Zaragoza.

In Asia, a subtropical steppe can be found in semi-arid lands that fringe the Thar Desert of the Indian subcontinent and the Badia of the Arabian peninsula.

In Australia, "subtropical steppe" can be found in a belt surrounding the most severe deserts of the continent and around the Musgrave Ranges.

In North America this environment is typical of transition areas between zones with a Mediterranean climate and true deserts, such as Reno, Nevada, the inner part of California, and much of western Texas and adjacent areas in Mexico.

Forest Steppe

A forest steppe is a temperate-climate ecotone and habitat type composed of grassland interspersed with areas of woodland or forest.

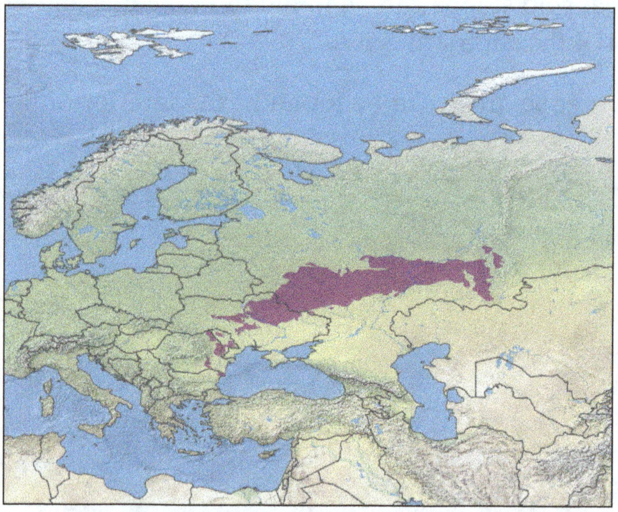

East European forest steppe (ecoregion PA0419).

Locations

Forest steppe primarily occurs in a belt of forest steppes across northern Eurasia from the eastern lowlands of Europe to eastern Siberia in northeast Asia. It forms transition ecoregions between the temperate grasslands and temperate broadleaf and mixed forests biomes. Much of Russia belongs to the forest steppe zone, stretches from Central Russia, across Volga, Ural, Siberian and Far East Russia.

Forest steppe landscape on the Volga Upland near the city of Saratov, Russia.

In upper North America another example of the forest steppe ecotone is the aspen parkland, in the central Prairie Provinces, northeastern British Columbia, and North Dakota. It is the transition ecoregion from the Great Plains prairie and steppe temperate grasslands to the Taiga biome forests in the north.

Devín forest steppe in Slovakia.

In central Asia the forest steppe ecotone is found in ecoregions in the mountains of the Iranian Plateau, in Iran, Afghanistan, and Balochistan.

Forest Steppe Ecoregions

- East European forest steppe forms a transition between the Central European and Sarmatic mixed forests to the north and the Pontic-Caspian steppe to the south. It extends from Romania in the west to the Ural Mountains in the east.
- The Kazakh forest steppe lies east of the Urals, between the West Siberian broadleaf and mixed forests and the Kazakh steppe.
- Altai montane forest and forest steppe.
- The Southern Siberian rainforest includes forest-steppe areas.
- Selenge-Orkhon forest steppe.
- The Daurian forest steppe lies between the Trans-Baikal conifer forests and East Siberian Taiga to the north and the Mongolian-Manchurian grassland to the south.
- Zagros Mountains forest steppe.
- Elburz Range forest steppe.
- Kopet Dag woodlands and forest steppe.
- Kuhrud-Kohbanan Mountains forest steppe.
- Canadian Aspen forests and parklands—North Dakota and Canada.

Alpine-steppe

The Alpine-steppe is a high altitude natural alpine grassland, which is a part of the Montane grasslands and shrublands biome.

Karakoram-West Tibetan Plateau alpine steppe.

Alpine-steppes are unique ecosystems found throughout the world, especially in Asia, where they make up 38.9% of the total Tibetan plateau grassland's area.

Characteristics

Alpine grasslands, like the Alpine-steppe, are characterized by their intense radiation, with direct solar radiation periods averaging 2916 hours annually. The average temperature in this ecosystem is very low. For example, they may experience temperatures around −10 °C in winter, and 10 °C in summer. Winters also tend to be long and cold, and summers are mild and short. This ecosystem also experiences year-long frost, with no reported frost-free season.

The annual rates of precipitation in Alpine-steppes are very low, with mean ranges falling anywhere between 280 and 300 mm. In addition, upwards to 80% of this falls between the months of May and September, causing the climate to be arid or semi-arid, making the environment much harsher for plant and livestock life.

Vegetation

Vegetation in the alpine steppe is very vulnerable to climate change. Average air temperature has been increasing by approximately 0.3 degrees Celsius every ten years since the 1960s. This is three times the global average, indicating the sensitivity of this area. Studies have been done that show that the spread of vegetation has changed dramatically since the Holocene period. The Tibetan Plateau is composed of three main regions, based on yearly precipitation levels and types of vegetation, namely the alpine meadow, alpine steppe, and the alpine desert-steppe. Since the Holocene, studies of fossil pollen records have shown that the alpine meadow has extended into areas that were previously alpine steppe as precipitation increased during that period. There is a unimodal pattern across precipitation and vegetation rain use efficiency (RUE), with an increasing trend in Alpine-steppe regions. RUE is lower here compared to the alpine meadow because of differences in species richness, soil texture, and soil carbon content.

Locoweed in Tibet.

Changes in vegetation have been used recently as an indicator of grassland degradation in the Tibetan Plateau, along with land desertification and decreased overall productivity. Vegetation shifts from non-poisonous to poisonous plants seem to correlate with increased land degradation. Plants defined as poisonous in the alpine grasslands area include species such as locoweed, which is known to be very invasive. Not only are poisonous plants an indicator of decline, they result in increased mortality of grazing animals. This invasion of poisonous species is spread across all regions of the Tibetan Plateau, but the alpine steppe is the most affected area.

The Tibetan Plateau is an extremely important area for livestock farming, and historically overgrazing has been as issue with regard to the sustainability of the vegetation in the area. Measures have been taken to regulate the use of these grasslands, including the implementation of protected or 'fenced areas'. While these measures are certainly a step in the right direction as far as sustainability legislation, they have not been shown to have a very strong effect on the above ground net primary productivity (ANPP).

Soil and Biome Composition

Based on studies done in the alpine steppe region of the Tibetan Plateau, different soil nutrients have differential effects on the nutrient composition and uptake of plants in the area. Soil phosphorus seems to have a much more significant impact on the nitrogen:phosphorus ratio in plants than soil nitrogen does. This type of finding can have implications for different nutrient conservation strategies among plant species in the same community, as plants seem to be more sensitive to changes in soil phosphorus than nitrogen, though nitrogen is still extremely important. What also makes this interesting is the fact that nitrogen is a limiting factor for plant growth, and so is actually critical for the overall health of the plant community. Grazing of herd animals has been shown to have a positive effect on the levels of nitrogen in the soil, though the return of nitrogen in the excrement. The addition of dung to the soils of this region in a laboratory setting resulted in increased availability of ammonia for plants (their primary nitrogen source). However, in an unaltered system, soil nitrogen tends to be more constant, whereas soil phosphorus is more influenced by climate variation, which may explain why, even though nitrogen is the limiting factor, phosphorus can be a greater influence on the N:P ratio of plant nutrients. alpine grassland temperature range from 14 degrees Fahrenheit in winter to 50 degrees Fahrenheit in summer.

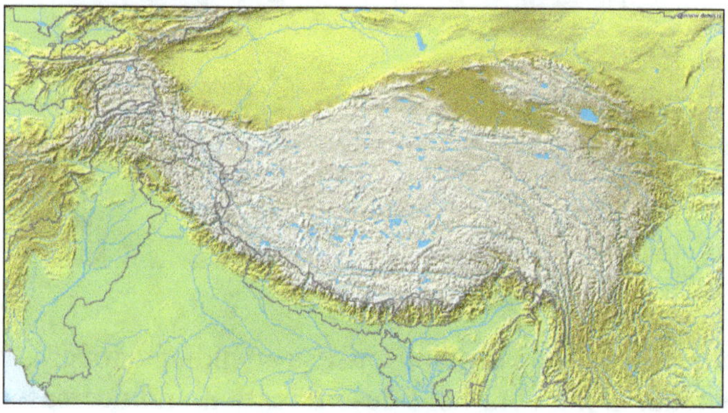

Topografic map of Tibetan Plateau.

Threats to Alpine-steppes

Yak in Tibet.

Because of their elevation, alpine regions are thought to experience higher rates of warming, causing them to be more sensitive and vulnerable to global climate change. Other major threats to Alpine-Steppes include overgrazing, as well as land use change associated with increases in population size. Because of this, the authorities in areas throughout China are under pressure to implement programs to protect and preserve this fragile ecosystem.

One such program is the "Retire Livestock and Restore Pastures" initiative, which requires the use of special enclosure fencing. The purpose of this protective fencing is to prevent the grazing activity of large livestock, like sheep, yaks, and goats, in an attempt to restore the degraded biomass, and maintain ecosystem function. Often these effects can best be seen by the changes they produce in the biogeochemical properties of the soil. The overall goal is to improve ecosystem carbon, nitrogen, and phosphorus storage, by increasing both vegetation and soil pools of these elements. This effect is crucial because even a small percent change in carbon storage can have a huge positive impact on atmospheric carbon dioxide and global carbon levels, as well as ecosystem sustainability. But carbon is not the only important factor. Low levels of nitrogen and phosphorus have also been found to limit plant growth and net primary productivity. In one study, exclusion fencing was found to increase the carbon stored in the biomass, as well as the nitrogen and phosphorus in the above-ground biomass. However this effect was minor, and not enough to compensate for the considerable loss of carbon, nitrogen, and phosphorus pools from the soil surface layer. Another study found exclusion fencing to be a beneficial tool in lowering carbon dioxide emissions, and increasing methane consumption, which improves both the soil carbon and nitrogen stores. Although findings are controversial, enclosure fencing remains a common practice in China because of the sensitivity of these grassland areas.

Artemisia capillaris.

At an elevation of 4,500–6,000 m, Northern Tibet's area is covered by approximately 94% grasslands, including the alpine-steppe and alpine meadow. The alpine-steppe in this area has less than 20% vegetation coverage, which consists mainly of Stipa purpurea, Artemisia capillaris Thunb, and Rhodiola rotundaia assemblages. Compared to the alpine meadow, the alpine-steppe is cooler, arid or semi-arid, with little precipitation and barren soils. The highest vegetation carbon pool can be found in August, and nitrogen and phosphorus concentrations in the area exhibit seasonal variations throughout the growing period.

Mammoth Steppe

Ukok Plateau, one of the last remnants of the mammoth steppe.

During the Last Glacial Maximum, the mammoth steppe was the Earth's most extensive biome. It spanned from Spain eastwards across Eurasia to Canada and from the arctic islands southwards to China. It had a cold, dry climate; the vegetation was dominated by palatable high-productivity grasses, herbs and willow shrubs, and the animal biomass was dominated by bison, horses, and the woolly mammoth. This ecosystem covered wide areas of the northern part of the globe, thrived for approximately 100,000 years without major changes, and then suddenly became all but extinct about 12,000 years ago.

The last glacial period, commonly referred to as the 'Ice Age', spanned from 126,000 YBP–11,700 YBP and was the most recent glacial period within the current ice age which occurred during the last years of the Pleistocene epoch. This arctic environment was very cold and dry and probably dusty, resembling mountaintop environments, and was very different from today's swampy tundra. It reached its peak during the last glacial maximum, when ice sheets commenced advancing from 33,000 years BP and reached their maximum positions 26,500 years BP. Deglaciation commenced in the Northern Hemisphere approximately 19,000 years BP, and in Antarctica approximately 14,500 years BP, which is consistent with evidence that it was the primary source for an abrupt rise in the sea level at that time.

During the peak of the last glacial maximum, a vast mammoth steppe stretched from the Iberian Peninsula across Eurasia and over the Bering land bridge into Alaska and the Yukon where it was stopped by the Wisconsin glaciation. This land bridge existed because more of the planet's water was locked up in ice than now and therefore the sea levels were lower. When the sea levels began to rise this bridge was inundated around 11,000 years BP.

During glacial periods, there is clear evidence for intense aridity due to water being held in glaciers and their associated effects on climate. The mammoth steppe was like a huge 'inner court' that was surrounded on all sides by moisture-blocking features: massive continental glaciers, high mountains, and frozen seas. These kept rainfall low and created more days with clear skies than are seen today, which increased evaporation in the summer leading to aridity, and radiation of warmth from the ground into the black night sky in the winter leading to cold. This is thought to have been caused by seven factors:

- The driving force for the core Asian steppe was an enormous and stable high-pressure system north of the Tibetan Plateau.
- Deflection of the larger portion of the Gulf Stream southward, past southern Spain onto the coast of Africa, reduced temperatures (hence moisture and cloud cover) that the North Atlantic Current brings to Western Europe.
- Growth of the Scandinavian ice sheet created a barrier to North Atlantic moisture.
- Icing over of the North Atlantic sea surface with reduced flow of moisture from the east.
- The winter (January) storm track seems to have swept across Eurasia on this axis.
- Lowered sea levels exposed a large continental shelf to the north and east producing a vast northern plain which increased the size of the continent to the north.
- North American glaciers shielded interior Alaska and the Yukon Territory from moisture flow. These physical barriers to moisture flow created a vast arid basin or protected 'inner court' spanning parts of three continents.

Biota

Climatic suitability for the woolly mammoths in the Late Pleistocene and Holocene. Increasing intensities of red represent increasing suitability of the climate and increasing intensities of green represent decreasing suitability. Black points are the records of mammoth presence for each of the periods. Black lines represent the northern limit of modern humans and black dotted lines indicate uncertainty in the limit of modern humans.

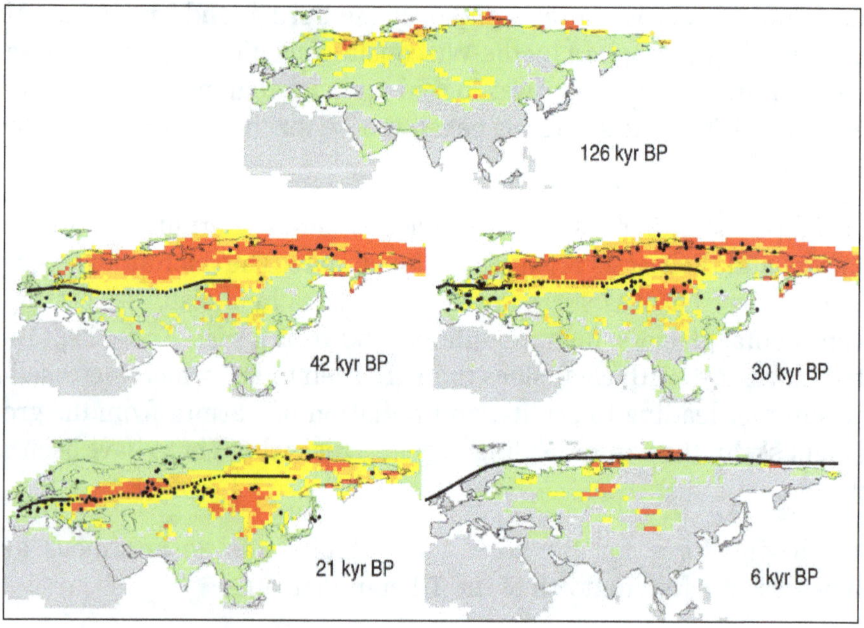

Animal biomass and plant productivity of the mammoth steppe were similar to today's African savannah. There is no comparison to it today.

Plants

Artemisia, Taymyr lowlands 24,000–10,300 YBP, Yakutia 22,500 YBP, Alaska and the Yukon 15,000-11,500 YBP.

Cyperaceae (sedges), Yakutia 22,500 YBP, Alaska and the Yukon 15,000-11,500 YBP.

The paleo-environment changed across time, a proposal that is supported from mammoth dung samples found in northern Yakutia. During Pleniglacial interstadials, alder, birch, and pine trees survived in northern Siberia, however during the Last Glacial Maximum only a treeless steppe vegetation existed. At the onset of the Late Glacial Interstadial (15,000–11,000 BP), global warming resulted in shrub and dwarf birch in northeastern Siberia, which was then colonized by open woodland with birch and spruce during the Younger Dryas (12,900–11,700 YBP). By the Holocene (10,000 YBP), patches of closed larch and pine forests developed. Past researchers had once assumed that the mammoth steppe was very unproductive because they had assumed that its soils had a very low carbon content; however, these soils (yedoma) were preserved in the permafrost of Siberia and Alaska and are the largest reservoir of organic carbon known. It was a highly productive environment. The vegetation was dominated by palatable high-productivity grasses, herbs and willow shrubs. Herbs were far more widespread than they are today, and were the main food source of the large plant eating mammals.

Gramineae (grasses), Taymyr lowlands 24,000–10,300 YBP, Yakutia 22,500 YBP, Alaska and the Yukon 15,000-11,500 YBP.

Potentilla (cinquefoil) Yakutia 22,500 YBP.

Animals

The mammoth steppe was dominated in biomass by bison, horse, and the woolly mammoth, and was the center for the evolution of the Pleistocene woolly fauna. On Wrangel Island, the remains of woolly mammoth, woolly rhinoceros, horse, bison and musk ox have been found. Reindeer and small animal remains do not preserve, but reindeer excrement has been found in sediment. In the most arid regions of the mammoth steppe that were to the south of Central Siberia and Mongolia, woolly rhinoceros were common but woolly mammoths were rare. Reindeer live in the far north of Mongolia today and historically their southern boundary passed through Germany and along the steppes of eastern Europe, indicating they once covered much of the mammoth steppe. Mammoths survived on the Taimyr Peninsula until the Holocene. A small population of mammoths survived on St. Paul Island, Alaska, up until 3750 BC, and the small mammoths of Wrangel Island survived until 1650 BC. Bison in Alaska and the Yukon, and horses and muskox in northern Siberia, have survived the loss of the mammoth steppe. One study has proposed that a change of suitable climate caused a significant drop in the mammoth population size, which made them vulnerable to hunting from expanding human populations. The coincidence of both of these impacts in the Holocene most likely set the place and time for the extinction of the woolly mammoth.

Decline of the Mammoth Steppe

The mammoth steppe had a cold, dry climate. During the past interglacial warmings, forests of trees and shrubs expanded northwards into the mammoth steppe, when northern Siberia, Alaska and the Yukon (Beringia) would have formed a mammoth steppe refugium. When the planet grew colder again, the mammoth steppe expanded. This ecosystem covered wide areas of the northern part of the globe, thrived for approximately 100,000 years without major changes, and then suddenly became extinct about 12,000 years ago.

There are two theories about the decline of the mammoth steppe.

Climate Change

The Climatic Hypothesis assumes that the vast mammoth ecosystem could have only existed within a certain range of climatic parameters. At the beginning of the Holocene 10,000 years ago, mossy forests, tundra, lakes and wetlands displaced mammoth steppe. It has been assumed that in contrast to other previous interglacials the cold dry climate switched to a warmer wetter climate that, in turn, caused the disappearance of the grasslands and their dependent megafauna.

The extinct steppe bison (Bison priscus) survived across the northern region of central eastern Siberia until 8000 years ago. A study of the frozen mummy of a steppe bison found in northern Yakutia, Russia indicated that it was a pasture grazer in a habitat that was becoming dominated by shrub and tundra vegetation. Higher temperature and rainfall led to a decrease in its previous habitat during the early Holocene, and this led to population fragmentation followed by extinction.

In 2017, a study looked at the environmental conditions across Europe, Siberia and the Americas from 25,000–10,000 YBP. The study found that prolonged warming events leading to deglaciation and maximum rainfall occurred just prior to the transformation of the rangelands that supported megaherbivores into widespread wetlands that supported herbivore-resistant plants. The study proposes that moisture-driven environmental change led to the megafaunal extinctions, and that Africa's trans-equatorial position allowed rangeland to continue to exist between the deserts and the central forests; therefore fewer megafauna species became extinct there.

Human Predation

The Ecosystem Hypothesis assumes that the vast mammoth ecosystem extended over a range of many regional climates and was not affected by climate fluctuations. Its highly productive grasslands were maintained by animals trampling any mosses and shrubs, and actively transpiring grasses and herbs dominated. At the beginning of the Holocene the rise in precipitation was accompanied by increased temperature, and so its climatic aridity did not change substantially. As a result of human hunting, the decreasing density of the animals was not enough to maintain the grasslands, leading to an increase in forests, shrubs and mosses with further animal reduction due to loss of feed. The mammoth continued to exist on isolated Wrangel Island until a few thousand years ago, and some of the other megafauna from that time still exist today, which indicates that something other than climate change was responsible for megafaunal extinctions.

Remains of mammoth that had been hunted by humans 45,000 YBP have been found at Yenisei Bay in the central Siberian Arctic. Two other sites in the Maksunuokha River valley to the south of the Shirokostan Peninsula, northeast Siberia, dated between 14,900 and 13,600 years ago showed the remains of mammoth hunting and the production of micro-blades similar to those found in northwest North America, suggesting a cultural connection.

Last Remnants

Ubsunur Hollow Biosphere Reserve located on the border of Mongolia and the Republic of Tuva is one of the last remnants of the mammoth steppe.

During the Holocene, the arid-adapted species became extinct or were reduced to minor habitats. Cold and dry conditions similar to the last glacial period are found today in the eastern Altai-Sayan mountains of Central Eurasia, with no significant changes occurring between the cold phase of the Pleistocene and the Holocene. Recent paleo-biome reconstruction and pollen analysis suggest that some present-day Altai-Sayan areas could be considered the closest analogy to the mammoth steppe environment. The environment of this region is considered to have been stable for the past 40,000 years. The Eastern part of the Altai-Sayan region forms a Last Glacial refugium. In both the Last Glacial and modern times, the eastern Altai-Sayan region has supported large herbivore and predator species adapted to the steppe, desert and alpine biomes where these biomes have not been separated by forest belts. None of the surviving Pleistocene mammals live in temperate forest, taiga, or tundra biomes. The areas of Ukok-Sailiugem in the southern Altai Republic, and Khar Us Nuur and Uvs Nuur (Ubsunur Hollow) in western Mongolia, have supported reindeer and saiga antelope since the glacial period.

Shrub-steppe

Shrub-steppe is the largest natural grassland in North America. It extends from southeastern Washington and eastern Oregon, through Idaho, Nevada, and Utah, and into western Wyoming and Colorado. Shrub refers to the most abundant plant species that grows in this ecoregion. "Steppe" is a Russian word that means a vast treeless plain. In the Mid-Columbia Basin, shrub-steppe winters are cold and wet with strong winds and blowing snow. Summers are hot and dry with temperatures that can reach above 100 degrees Fahrenheit during the day, then cool at night. Here, less than 7 inches of rain falls per year.

The Rattlesnake Hills in the Columbia Basin are part of the shrub-steppe ecoregion.

The kind of plants that grow in a region is determined by many things, including climate and soil, geology and topography. Under natural conditions, shrub-steppe lands are covered with grasses and shrubs. The most common shrub, or woody plant, is big sagebrush. Other shrubs you probably have seen in our area, but may not have recognized, are rabbitbrush, greasewood, hopsage, bitterbrush, and buckwheat. They grow in communities with grasses such as bluebunch wheatgrass and Sandberg's bluegrass.

Rabbitbrush often is mistaken for sagebrush because both shrubs are medium sized, gray-green, frequently grow together, and bloom in late summer or fall. One way you can distinguish between them is to sniff a leaf. Rabbitbrush doesn't smell as strong as sage, and it has long, unlobed leaves. Sagebrush has three lobes at the tip each leaf. Also, rabbitbrush has large bright yellow flowers, while the flowers of sagebrush are tiny and inconspicuous.

Although identifying shrubs may seem confusing, you probably are familiar with some shrub-steppe wildflowers. Likely you have seen the bright yellow flowers and heart-shaped leaves of Carey's balsamroot covering Mid-Columbia hillsides in spring, along with rich displays of purple-blue lupine. In sandy areas, you also may have seen long-leaf phlox climbing up through the branches of sagebrush.

The shrub-steppe ecoregion supports a variety of birds, mammals, and reptiles/amphibians. More than 200 bird species and 30 kinds of mammals are known to live in our arid region. The types of plants that grow in the shrub-steppe determine the number and kinds of wildlife that can live here too. Wildlife depend on plants for food, concealment, protective cover, and for some species of birds, for nesting.

Some breeding birds that live in the Mid-Columbia are chukar, common ravens, black-billed magpies, Swainson's hawks, red-tailed hawks, horned larks, western meadowlarks, and burrowing owls. Mammals that make their home here include such species as elk, mule deer, bobcat, and black-tailed jackrabbit. The Great Basin pocket mouse and Townsend's ground squirrel also live here. These small mammals provide a source of food for badgers, coyotes, and nesting hawks. Less common in our area are reptiles and amphibians, though we do have side-blotched lizards and northern Pacific Rattlesnakes. Great Basin spadefoot toads have been seen near Rattlesnake Springs in the Rattlesnake Hills.

References

- Ecosystem: newworldencyclopedia.org, Retrieved 01 May, 2020
- Threats-to-ecosystems: enviropol.com, Retrieved 28 June, 2020
- Forest, science: britannica.com, Retrieved 04 August, 2020
- Barange M, Field JG, Harris RP, Eileen E, Hofmann EE, Perry RI and Werner F (2010) Marine Ecosystems and Global Change Oxford University Press. ISBN 978-0-19-955802-5
- Forest-ecosystem-types-characteristic-features-and-structure, environment-ecosystem- 28240: yourarticlelibrary.com, Retrieved 16 May, 2020
- Shrubsteppe: workbasedlearning.pnnl.gov, Retrieved 23 June, 2020
- Oberndorfer, E. C.; Lundholm, J. T. (2008). "Species richness, abundance, rarity and environmental gradients in coastal barren vegetation". Biodiversity and Conservation. 18 (6): 1523–1553. doi:10.1007/s10531-008-9539-5

3

Understanding Ecosystem Management

The process of management that is aimed at conserving ecological services and restoring natural resources is termed as ecosystem management. Conservation biology, soil conservation, watershed management, etc. are some of the areas studied under its domain. All these aspects related to ecosystem management have been carefully analyzed in this chapter.

Ecosystem Approach

The Ecosystem Approach is a concept that integrates the management of land, water and living resources and aims to reach a balance between three objectives: conservation of biodiversity; its sustainable use; and equitable sharing of benefits arising from the utilisation of natural resources. It is the primary implementation framework of the Convention on Biological Diversity (CBD).

An Ecosystem Approach takes into account that humans and cultural diversity are an integral element of most ecosystems. It applies appropriate scientific methodologies, focused on various levels of biological organisation, which encompass the fundamental structure, processes, functions and interactions amongst and between organisms and their environment.

The Ecosystem Approach is not a formula, but a framework that can be adapted to suit various issues and situations. The definition of an Ecosystem Approach does not specify any particular spatial unit or scale; therefore it can refer to any ecological unit at any scale.

It is important to recognise that the Ecosystem Approach does not provide an all encompassing solution as its application depends upon local, provincial, national, regional or global conditions. An Ecosystem Approach should not be regarded as a strategy that supplants other techniques and tools; where ever possible existing strategies and methodologies should be used in conjunction to address complex problems and issues.

The Ecosystem Approach is an adaptive management strategy that can be employed to deal with the complex and dynamic nature of ecosystems and counteract the lack of knowledge or comprehension of their functioning. Ecosystem processes are often non-linear, fluctuate spatially and temporally and frequently show time-lags; these discontinuities can create a high level of uncertainty which the Ecosystem Approach can help overcome.

Adopting this balanced approach ensures that natural resources and society as a whole are positioned in the centre of the decision making process, ensuring a more equitable and long-term future is tenable.

Mangroves are an integral part of ecosystems.

Ecosystem management is a process that aims to conserve major ecological services and restore natural resources while meeting the socioeconomic, political, and cultural needs of current and future generations. The principal objective of ecosystem management is the efficient maintenance and ethical use of natural resources. It is a multifaceted and holistic approach which requires a significant change in how the natural and human environments are identified.

Several approaches to effective ecosystem management engage conservation efforts at both local and landscape levels and involve: adaptive management, natural resource management, strategic management, and command and control management.

A variety of definitions exist: Robert T. Lackey defined ecosystem management as "the application of ecological and social information, options, and constraints to achieve desired social benefits within a defined geographic area and over a specified period." F. Stuart Chapin and coauthors define it as "the application of ecological science to resource management to promote long-term sustainability of ecosystems and the delivery of essential ecosystem goods and services," while Norman Christensen and coauthors defined it as "management driven by explicit goals, executed by policies, protocols, and practices, and made adaptable by monitoring and research based on our best understanding of the ecological interactions and processes necessary to sustain ecosystem structure and function." Peter Brussard and colleagues defined it as "managing areas at various scales in such a way that ecosystem services and biological resources are preserved while appropriate human use and options for livelihood are sustained".

The definitions of ecosystem management are typically vague. Several core principles define and bound the concept and provide operational meaning:

- Ecosystem management reflects a stage in the continuing evolution of social values and priorities; it is neither a beginning nor an end.
- Ecosystem management is place-based and the boundaries of the place must be clearly and formally defined.
- Ecosystem management should maintain ecosystems in the appropriate condition to achieve desired social benefits.
- Ecosystem management should take advantage of the ability of ecosystems to respond to a variety of natural and man-made stressors, but all ecosystems have limited ability to accommodate stressors and maintain a desired state.
- Ecosystem management may or may not result in emphasis on biological diversity.
- The term sustainability, if used at all in ecosystem management, should be clearly defined specifically, the time frame of concern, the benefits and costs of concern, and the relative priority of the benefits and costs.
- Scientific information is important for effective ecosystem management, but is only one element in a decision-making process that is fundamentally one of public and private choice.

A fundamental principle is the long-term sustainability of the production of goods and services by the ecosystem; "intergenerational sustainability is a precondition for management, not an afterthought". It also requires clear goals with respect to future trajectories and behaviors of the system being managed. Other important requirements include a sound ecological understanding of the system, including connectedness, ecological dynamics and the context in which the system is embedded. An understanding of the role of humans as components of the ecosystems and the use of adaptive management is also important. While ecosystem management can be used as part of a plan for wilderness conservation, it can also be used in intensively managed ecosystems (for example, agroecosystem and close to nature forestry).

As a concept of natural resource management, ecosystem management remains both ambiguous and controversial, in part because some of its formulations rest on policy and scientific assertions that are contested. These assertions are important to understanding much of the conflict surrounding ecosystem management. Professional natural resource managers, typically operating from within government bureaucracies and professional organizations, often mask debate over controversial assertions by depicting ecosystem management as an evolution of past management approaches.

Stakeholders

Stakeholders are individuals or groups of people who are affected by environmental decisions and actions, but they also may have power to influence the outcomes of environmental decisions relating to ecosystem management. The complex nature of decisions made in ecosystem management, from local to international scales, requires stakeholder participation with a diversity of knowledge, perceptions, and values of nature. Stakeholders will often have different interests in ecosystem services. This means effective management of ecosystems requires a negotiation process that develops mutual trust in issues of common interest with the objective of creating mutually beneficial partnerships.

Adaptive Management

Adaptive management is based on the concept that predicting future influences/disturbance to an ecosystem is limited and unclear. Therefore, the goal of adaptive management is to manage the ecosystem so it maintains the greatest amount of ecological integrity, but also to utilize management practices that have the ability to change based on new experience and insights.

Adaptive management aims to identify uncertainties in the management of an ecosystem while using hypothesis testing to further understand the system. In this regard, adaptive management encourages learning from the outcomes of previously implemented management strategies. Ecosystem managers form hypotheses about the ecosystem and its functionality and then implement different management techniques to test the hypotheses. The implemented techniques are then analyzed to evaluate any regressions or improvements in functionality of the ecosystem caused by the technique. Further analysis allows for modification of the technique until it successfully meets the ecological needs of the ecosystem. Thus, adaptive management serves as a "learning by doing" method for ecosystem management.

Adaptive management has had mixed success in the field of ecosystem management, possibly because ecosystem managers may not be equipped with the decision-making skills needed to undertake an adaptive management methodology. Additionally, economic, social and political priorities can interfere with adaptive management decisions. For this reason, adaptive management should be a social process as well as scientific, focusing on institutional strategies while implementing experimental management techniques.

Natural Resource Management

The term natural resource management is frequently used when dealing with a particular resource for human use rather than managing the whole ecosystem. A main objective of natural resources management is the sustainability for future generations, which appoints ecosystem managers to balance natural resources exploitation and conservation over long-term timeframe. The balanced relationship of each resource in an ecosystem is subject to change at different spatial and temporal scales. Dimensions such as watersheds, soils, flora, and fauna need to be considered individually and on a landscape level. A variety of natural resources are utilized for food, medicine, energy and shelter.

The ecosystem management concept is based on the relationship between sustainable resource maintenance and human demand for use of natural resources. Therefore, socioeconomics factors significantly affect natural resource management. The goal of a natural resource manager is to fulfill the demand for a given resource without causing harm to the ecosystem, or jeopardizing the future of the resource. Partnerships between ecosystem managers, natural resource managers and stakeholders should be encouraged in order to promote a more sustainable use of limited natural resources. Natural resource managers must initially measure the overall integrity of the ecosystem they are involved in. If the ecosystem supporting resources is healthy, managers can decide on the ideal amount of resource extraction, while leaving enough to allow the resource to replenish itself for subsequent harvests. Historically, some natural resources have experienced limited human disturbance and therefore have been able to subsist naturally. However, some ecosystems such as forests, which typically provide considerable timber resources; have sometimes undergone successful reforestation processes and consequently have accommodated the needs of future generations. A successfully managed resource will provide for current demand while leaving enough to repopulate and provide for future demand.

Human populations have been increasing rapidly, introducing new stressors to ecosystems, such as climate change and influxes of invasive species. As a result, the demand for natural resources is unpredictable. Although ecosystem changes may occur gradually, the cumulative changes can have negative effects for humans and wildlife. Geographic Information Systems (GIS) and Remote Sensing Applications can be used to monitor and evaluate natural resources by mapping them in local and global scales. These tools will continue to be highly beneficial in natural resources management.

Strategic Management

Strategic management encourages the establishment of goals that will benefit the ecosystem while keeping socioeconomic and politically relevant issues in mind. Strategic management differs from other types of ecosystem management because it keeps stakeholders involved and relies on their input to develop the best management strategy for an ecosystem. Similarly to other modes of ecosystem management, this method places a high level of importance on evaluating and reviewing any changes, progress, or negative impacts and prioritizes flexibility in adapting management protocols as a result of new information.

Landscape Level Conservation

Landscape level conservation is a method that considers wildlife needs at a broader landscape level scale when implementing conservation initiatives. This approach to ecosystem management involves the consideration of broad scale interconnected ecological systems that acknowledges the whole scope of an environmental problem. In a human-dominated world, weighing the landscape requirements of wildlife versus the needs of humans is a complicated matter.

Landscape level conservation is carried out in a number of ways. A wildlife corridor, for example, is a connection between otherwise isolated habitat patches that are proposed as a solution to habitat fragmentation. In some landscape level conservation approaches, a key species vulnerable to landscape alteration is identified and its habitat requirements are assessed in order to identify the best option for protecting their ecosystem. However, lining up the habitat requirements of numerous species in an ecosystem can be difficult, which is why more comprehensive approaches to further understand these variations have been considered in landscape level conservation.

Human-induced environmental degradation is an increasing problem globally, which is why landscape level ecology plays an important role in ecosystem management. Traditional conservation methods targeted at individual species need to be modified to include the maintenance of wildlife habitats through consideration of both human-induced and natural environmental factors.

Command and Control Management

Command and control management utilizes a linear problem solving approach where a perceived problem is solved through controlling devices such as laws, threats, contracts and/or agreements. This top-down approach is used across many disciplines and works best with problems that are relatively simple, well-defined and work in terms of cause and effect. The application of command and control management has often attempted to control nature in order to improve product extractions, establish predictability and reduce threats. Some obvious examples of command and control management actions include: the use of herbicides and pesticides to safeguard crops in order to harvest more products; the culling of predators in order to obtain larger, more reliable game species; and the safeguarding of timber supply, by suppressing forest fires.

Attempts at command and control management often backfire (a literal problem in forests that have been protected from fire by humans and are subsequently full of fuel build-up) in ecosystems due to their inherent complexities. Consequently, there has been a transition away from command and control management due to many undesirable outcomes and a stronger focus has been placed on more holistic approaches that focus on adaptive management and finding solutions through partnerships.

Ecosystem-based Management

Ecosystem-based management is an environmental management approach that recognizes the full array of interactions within an ecosystem, including humans, rather than considering single issues, species, or ecosystem services in isolation. It can be applied to studies in the terrestrial and marine environments with challenges being attributed to both. In the marine realm, they are highly challenging to quantify due to highly migratory species as well as rapidly changing environmental and anthropogenic factors that can alter the habitat rather quickly. To be able to manage fisheries efficiently and effectively it has become increasingly more pertinent to understand not only the biological aspects of the species being studied, but also the environmental variables they are experiencing. Population abundance and structure, life history traits, competition with other species, where the stock is in the local food web, tidal fluctuations, salinity patterns and anthropogenic influences are among the variables that must be taken into account to fully understand the implementation of a "ecosystem-based management" approach. Interest in ecosystem-based management in the marine realm has developed more recently, in response to increasing recognition of the declining state of fisheries and ocean ecosystems. However, due to a lack of a clear definition and the diversity involved with the environment, the implementation has been lagging.

Terrestrial ecosystem-based management (often referred to as ecosystem management) came into its own during the conflicts over endangered species protection (particularly the northern spotted owl), land conservation, and water, grazing and timber rights in the western United States in the 1980s and 1990s.

Implementation

Because ecosystem-based management is applied to large, diverse areas encompassing an array of interactions between species, ecosystem components, and humans, it is often perceived as a complex process that is difficult to implement. Slocombe also noted that in addition, uncertainty is common and predictions are difficult. However, in light of significant ecosystem degradation, there is a need for a holistic approach that combines environmental knowledge and co-ordination with governing agencies to initiate, sustain and enforce habitat and species protection, and include public education and involvement. As a result, ecosystem-based management will likely be increasingly used in the future as a form of environmental management. Some suggestions for implementing ecosystem-based management and what the process may involve are as follows.

Goals and Objectives

Defining clear and concise goals for ecosystem-based management is one of the most important steps in effective ecosystem-based management implementation. Goals must move beyond science-based or science-defined objectives to include social, cultural, economic and environmental importance. Of equal importance is to make sure that the community and stake-holders are involved throughout the entire process. A single, end-all goal cannot be the solution, but instead a combination of goals and their relationships with each other should be the focus.

Goals should be broadly applicable, measurable and readily observable, and ideally be collectively supported in order to be achievable. The idea is to provide direction for both thinking and action and should try to minimize managing ecosystems in a static state. Goals should also be flexible enough to incorporate a measure of uncertainty and be able to evolve as conditions and knowledge change. This may involve focusing on specific threatening processes, such as habitat loss or introduced invasive species, occurring within an ecosystem. Overall the goals should be integrative, to include the structure, organization and processes of the management of an area. Correct ecosystem-based management should be based in goals that are both "substantive", to explain the aims and importance of protecting an area, and "procedural", to explain how substantive goals will be met.

Some steps of ecosystem-based management may include:

Scoping

This step involves the acquisition of data and knowledge from various sources in order to provide a thorough understanding of critical ecosystem components. Sources may include literature, informal sources such as aboriginal residents, resource users, and/or environmental experts. Data may also be gained through statistical analyses, simulation models, or conceptual models.

Defining Indicators

Ecological indicators are useful for tracking or monitoring an ecosystem's status and can provide feedback on management progress. Examples may include the population size of a species or the levels of toxin present in a body of water. Social indicators may also be used such as the number or types of jobs within the environmental sector or the livelihood of specific social groups such as indigenous peoples.

Setting Thresholds

It is suggested to set-up thresholds for each indicator and setting targets that would represent a desired level of health for the ecosystem. Examples may include species composition within an ecosystem or the state of habitat conditions based on local observations or stakeholder interviews. Thresholds can be used to help guide management, particularly for a species by looking at the conservation status criteria established by either state or federal agencies and using models such as the minimum viable population size.

Risk Analysis

A range of threats and disturbances, both natural and human, often can affect indicators. Risk is defined as the sensitivity of an indicator to an ecological disturbance. Several models can be used to assess risk such as population viability analysis.

Monitoring

Evaluating the effectiveness of the implemented management strategies is very important in determining how management actions are affecting the ecosystem indicators. Evaluation: This final step involves monitoring and assessing data to see how well the management strategies chosen are performing relative to the initial objectives stated. The use of simulation models or multi-stakeholder groups can help to assess management.

It is important to note that many of these steps for implementing ecosystem-based management are limited by the governance in place for a region, the data available for assessing ecosystem status and reflecting on the changes occurring, and the time frame in which to operate.

Challenges

Because ecosystems differ greatly and express varying degrees of vulnerability, it is difficult to apply a functional framework that can be universally applied. These outlined steps or components of ecosystem-based management can, for the most part, be applied to multiple situations and are only suggestions for improving or guiding the challenges involved with managing complex issues. Because of the greater amount of influences, impacts, and interactions to account for, problems, obstacles and criticism often arise within ecosystem-based management. There is also a need for more data, spatially and temporally to help management make sound decisions for the sustainability of the stock being studied.

The first commonly defined challenge is the need for meaningful and appropriate management units. These units must be broad and contain value for people in and outside of the protected area. For example, Aberley suggests the use of "bioregions" as management units, which can allow peoples involvement with that region to come through. To define management units as inclusive regions rather that exclusive ecological zones would prevent further limitations created by narrow or restricting political and economic policy created from the units. Slocombe suggests that better management units should be flexible and build from existing units and that the biggest challenge is creating truly effect units for managers to compare against.

Another issue is in the creation of administrative bodies. They should operate as the essence of ecosystem-based management, working together towards mutually agreed upon goals. Gaps in administration or research, competing objectives or priorities between management agencies and governments due to overlapping jurisdictions, or obscure goals such as sustainability, ecosystem integrity, or biodiversity can often result in fragmented or weak management. In addition, Tallis stated that limited knowledge of ecosystem components and function and time constraints that can often limit objectives to only those that can be addressed in the short-term.

The most challenging issue facing ecosystem-based management is that there exists little knowledge about the system and its effectiveness. Slocombe stated that with limited resources available on how to implement the system it is hard to find support for its use.

Criticism of ecosystem-based management include its reliance on analogy and comparisons, too broadly applied frameworks, its overlap with or duplication of other methods such as ecosystem management, environmental management, or integrated ecosystem assessment, its vagueness in concepts and application, and its tendency to ignore historical, evolutionary or individual factors that may heavily influence ecosystem functioning.

Tallis stated that ecosystem-based management is seen as a critical planning and management framework for conserving or restoring ecosystems though it is still not widely implemented. An ecosystem approach addresses many relationships across spatial, biological, and organizational scales and is a goal-driven approach to restoring and sustaining ecosystems and functions. In addition, ecosystem-based management involves community influence as well as planning and management from local, regional and national government bodies and management agencies. All must be in collaboration in order to develop a desired future of ecosystem conditions, particularly where ecosystems have undergone radical degradation and change. To move forward, ecosystem-based management should be approached through adaptive management, allowing flexibility and inclusiveness to deal with constant environmental, societal, and political change.

Marine Systems

Ecosystem-based management for marine environments moves away from the traditional strategies in which single species and single sectors are managed individually; rather it is an integrated approach which considers all key activities, particularly anthropogenic, that affect marine environments. The management must take into account the life history of the fish being studied, its association with the surrounding environment, its place in the food web, where it prefers to reside in the water column and how it is affected by human pressures. The objective is to ensure sustainable ecosystems, thus protecting the resources and services they provide for future generations.

In recent years, there has been increasing recognition of disruption to marine ecosystems resulting from climate change, overfishing, nutrient and chemical pollution from land runoff, coastal development, bycatch, habitat destruction and other anthropogenic sources. There are very clear links between human activities and marine ecosystem functioning; this has become an issue of high importance because there are many services provided by marine ecosystems that are declining as a result of these impacts. These services include the provision of food, fuel, mineral resources, pharmaceuticals, as well as opportunities for recreation, trade, research and education.

There is an urgent need to improve the management of these declining ecosystems, particularly in coastal areas, to ensure a sustainable future. Human communities depend on marine ecosystems for important resources, but without holistic management these ecosystems are likely to collapse. It has been suggested that the degradation of marine ecosystems is largely the result of poor governance and that new approaches to management are required. The Pew Oceans Commission and the United States Commission on Ocean Policy have indicated the importance of moving from current piecemeal management to a more integrated ecosystem-based approach.

Stock Assessment

In marine environments, it can be increasingly difficult to determine what part of the population a certain stock comes from. In the case of the Southern bluefin tuna, it is all one taxonomic stock while in the Pacific Northwest, each river has its own stock of pacific salmon and thus, the clear definition of a particular stock can be quite challenging. With knowledge on the specific species life history, stock assessment can be much easier and therefore, knowledge on the animals breeding patterns, juvenile state and morphological patterns are essential.

Dead salmon in spawning season.

Bottom Up or Top Down

When trying to analyze the specific species, the manager needs to fully understand where the species is in terms of the food web. In the North Sea, the puffin population crashed and no one was quite sure why. After further analysis, they realized that the decline in seabird numbers was due to the sandeel population being severely depleted due to fishing pressures and this is an example of top down. In Alaska, the sea otter was hunted mainly for its pelt and was nearly pushed to extinction. As their numbers dwindled, the kelp beds were disappearing as well at an alarming rate due to the sea urchins having no predators and eating all of the kelp. As the sea otters were removed, the rest of the ecosystem in the food chain beneath them were affected, an example of top-down.

Post-dip pose.

Bycatch

In the Gulf of Mexico, the red snapper is one of the most important economical species there is. However, it is a very difficult species to manage due to the bycatch associated with the shrimping industry. The juveniles are caught in the smaller mesh size used for the shrimp and thus, a large portion of the red snapper population is subject due to mortality not associated with the red snapper fishery.

Key Elements

Connections

At its core, ecosystem-based management is about acknowledging interdependency connections, including the linkages between marine ecosystems and human societies, economies and institutional systems, as well as those among various species within an ecosystem and among ocean places that are linked by the movement of species, materials, and ocean currents. Of particular importance is how these factors all react and involve each other. In the Caribbean, the spiny lobster is managed based on a classic population model that for most fishery species works quite well. However, this species will grow and then halt its growth when it need to molt its shell and thus instead of a continuous growth cycle, it will pause its growth and invest its energy in a new shell. To further complicate matters, it slows this process down as it gets older to invest more energy into reproduction thus further deviating itself from the von Bertalanffy model of growth that was applied to it. The more information we can gather about an ecosystem and all of the interconnected factors which affect it, the more capable we will be of better managing that system.

Cumulative Impacts

Ecosystem-based management focuses on how individual actions affect the ecosystem services that flow from coupled socio-ecological systems in an integrated fashion, rather than considering these impacts in a piecemeal manner. Loss of biodiversity in marine ecosystems is an example of how cumulative effects from different sectors can impact on an ecosystem in a compounding way. Overfishing, coastal development, filling and dredging, mining and other human activities all contribute to the loss of biodiversity and therefore degradation of the ecosystem. Work is needed prior to the carrying out of the research to understand the total effects that each species can have on each other and also on the environment. It must be carried out every year as well as species are changing their life history traits and their relationship with the environment as humans are continually modifying the environment.

Interactions between Sectors

The only way to deal with the cumulative effects of human influences on marine ecosystems is for various contributing sectors to set common goals for the protection or management of ecosystems. While some policies may only affect a single sector, others may affect multiple sectors. A policy for the protection of endangered marine species, for example, could affect recreational and commercial fisheries, mining, shipping and tourism sectors to name a few. More effective ecosystem management would result from the collective adoption of policies by all sectors, rather than each sector creating their own isolated policies. For example, in the Gulf of Mexico there are oil rigs, recreational fisheries, commercial fisheries and multiple tourist attractions. One of the main fisheries is that of the Red Snapper which inhabits much of the Gulf and employs thousands of people in the commercial and recreational fishery. During the Deepwater oil spill it became abundantly obvious that it negatively affected the population numbers as well as the integrity of the catch that was being made. The species not only suffered higher mortality rates but the market was less trusting of the product. An environmental disaster interacted with the commercial, recreational, and economic sector for a specific species.

Changing Public Perceptions

Not all members of the public will be properly informed, or be fully aware, of current threats to marine ecosystems and it is therefore important to change public perceptions by informing people about these issues. It is important to consider the interest of the public when making decisions about ocean management and not just those who have a material interest because community support is needed by management agencies in order to make decisions. The Great Barrier Reef Marine Park Authority (GBRMPA) faced the issue of poor public awareness in their proposed management strategy which included no-take fishing zones. Olssen addressed this problem by starting a 'reef under pressure' information campaign to prove to the public that the Great Barrier Reef is under threat from human disturbances, and in doing so were successful in gaining public support.

Bridging Science and Policy

To ensure that all key players are on the same page, it is important to have communication between managers, resource users, scientists, government bodies and other stakeholders. Leslie and McLeod stated that proper engagement between these groups will enable the development of management initiatives that are realistic and enforceable as well as effective for ecosystem management. If certain small-scale players are not involved or informed, it is highly unlikely and equally challenging to get them to cooperate as well as to follow the rules that need to be put in place. It is of the utmost importance to have every stakeholder involved with every step of the process to increase the cohesion of the process.

Embracing Change

Coupled social-ecological systems are constantly changing in ways that cannot be fully predicted or controlled. Understanding the resilience of ecosystems, i.e. the extent to which they can maintain structure, function, and identity in the face of disturbance, can enable better prediction of how ecosystems will respond to both natural and anthropogenic perturbations, and to changes in environmental management. With how much modification humans are doing to environments, it is important to understand these changes on a yearly basis as well. Some species are changing their life histories, Flounder, due to the increased pressures that humans are placing on the environment. Thus, when a manager or government does an assessment on the ecosystem for a given year, the relationship that a species has to others can change very quickly and thus negate the model that you use for an ecosystem very quickly if not redefined.

Multiple Objectives

Ecosystem-based management focuses on the diverse benefits provided by marine systems, rather than on single ecosystem services. Such benefits or services include vibrant commercial and recreational fisheries, biodiversity conservation, renewable energy from wind or waves and coastal protection. The goal is to provide a sustainable fisheries while incorporating the impacts of other aspects on that resource. When managed correctly, an ecosystem-based model can greatly improve not only the resource being managed, but those associated with it.

Learning and Adaptation

Because of the lack of control and predictability of coupled social-ecological systems, an adaptive management approach is recommended. There can be multiple different factors that must be overcome (fisheries, pollution, borders, multiple agencies, etc.) to create a positive outcome. Managers must be able to react and adapt as to limit the variance associated with the outcome.

Other Examples

Great Bear Rainforest

The Land and Resource Management Planning (LRMP) was implemented by the British Columbia Government (Canada) in the mid-1990s in the Great Bear Rainforest in order to establish a multiparty land-use planning system. The aim was to "maintain the ecological integrity of terrestrial, marine and freshwater ecosystems and achieve high levels of human well-being". The steps described in the programme included: protect old-growth forests, maintain forest structure at the stand level, protect threatened and endangered species and ecosystems, protect wetlands and apply adaptive management. MacKinnon highlighted that the main limitation of this program was the social and economic aspects related to the lack of orientation to improve human well-being.

The Great Lakes

A Remedial Action Plan (RAP) was created during the Great Lakes Water Quality Agreement that implemented ecosystem-based management. The transition, from "a narrow to a broader approach " was not easy because it required the cooperation of both the Canadian and American governments. This meant different cultural, political and regulatory perspectives were involved with regards to the lakes. Eight principles required to make the implementation of ecosystem-based management efficacious: "broad-based stakeholder involvement; commitment of top leaders; agreement on information needs and interpretation; action planning within a strategic framework; human resource development; results and indicators to measure progress; systematic review and feedback; and stakeholder satisfaction".

Scallop Aquaculture in Sechura Bay

Peruvian Bay Scallop is grown in the benthic environment. Intensity of the fishery has caused concern over recent years and there has been a shift to more of an environmental management scheme. They are now using food web models to assess the current situation and to calibrate the stocking levels that are needed. The impacts of the scallops on the ecosystem and on other species are now being taken into account as to limit phytoplankton blooms, overstocking, diseases and overconsumption in a given year. This study is proposed to help guide both fisherman and managers in their goal of providing long term success for the fishery as well as the ecosystem they are utilizing.

Dam Removal

The Elwha dam removal in Washington state is the largest dam removal project in the United States. Not only was it blocking several species of salmon from reaching their natural habitat, it also had millions of tons of sediment built up behind it.

Environmental Resource Management

The shrinking Aral Sea, an example of poor water resource management diverted for irrigation.

Environmental resource management is the management of the interaction and impact of human societies on the environment. It is not, as the phrase might suggest, the management of the environment itself. Environmental resources management aims to ensure that ecosystem services are protected and maintained for future human generations, and also maintain ecosystem integrity through considering ethical, economic, and scientific (ecological) variables. Environmental resource management tries to identify factors affected by conflicts that rise between meeting needs and protecting resources. It is thus linked to environmental protection, sustainability and integrated landscape management.

Significance

Environmental resource management is an issue of increasing concern, as reflected in its prevalence in several texts influencing global sociopolitical frameworks such as the Brundtland Commission's Our Common Future, which highlighted the integrated nature of environment and international development and the Worldwatch Institute's annual State of the World reports.

The environment determines the nature of people, animals, plants, and places around the Earth, affecting behaviour, religion, culture and economic practices.

Scope

Improved agricultural practices such as these terraces in northwest Iowa can serve to preserve soil and improve water quality.

Environmental resource management can be viewed from a variety of perspectives. It involves the management of all components of the biophysical environment, both living (biotic) and non-living (abiotic), and the relationships among all living species and their habitats. The environment also involves the relationships of the human environment, such as the social, cultural and economic environment, with the biophysical environment. The essential aspects of environmental resource management are ethical, economical, social, and technological. These underlie principles and help make decisions.

The concept of environmental determinism, probabilism and possibilism are significant in the concept of environmental resource management.

Environmental resource management covers many areas in science, including geography, biology, social sciences, political sciences, public policy, ecology, physics, chemistry, sociology, psychology, and physiology.

Aspects

Ethical

Environmental resource management strategies are intrinsically driven by conceptions of human-nature relationships. Ethical aspects involve the cultural and social issues relating to the environment, and dealing with changes to it. All human activities take place in the context of certain types of relationships between society and the bio-physical world (the rest of nature), and so, there is a great significance in understanding the ethical values of different groups around the world. Broadly speaking, two schools of thought exist in environmental ethics: Anthropocentrism and Ecocentrism, each influencing a broad spectrum of environmental resource management styles along a continuum. These styles perceive different evidence, imperatives, and problems, and prescribe different solutions, strategies, technologies, roles for economic sectors, culture, governments, and ethics, etc.

Anthropocentrism

Anthropocentrism, "an inclination to evaluate reality exclusively in terms of human values" is an ethic reflected in the major interpretations of Western religions and the dominant economic paradigms of the industrialised world. Anthropocentrism looks at nature as existing solely for the benefit of humans, and as a commodity to use for the good of humanity and to improve human quality of life. Anthropocentric environmental resource management is therefore not the conservation of the environment solely for the environment's sake, but rather the conservation of the environment, and ecosystem structure, for humans sake.

Ecocentrism

Ecocentrists believe in the intrinsic value of nature while maintaining that human beings must use and even exploit nature to survive and live. It is this fine ethical line that ecocentrists navigate between fair use and abuse. At an extreme of the ethical scale, ecocentrism includes philosophies such as ecofeminism and deep ecology, which evolved as a reaction to dominant anthropocentric paradigms. In its current form, it is an attempt to synthesize many old and some new philosophical attitudes about the relationship between nature and human activity, with particular emphasis on ethical, social, and spiritual aspects that have been downplayed in the dominant economic worldview.

Economics

A water harvesting system collects rainwater from the Rock of Gibraltar into pipes that lead to tanks excavated inside the rock.

The economy functions within, and is dependent upon goods and services provided by natural ecosystems. The role of the environment is recognized in both classical economics and neoclassical economics theories, yet the environment was a lower priority in economic policies from 1950 to 1980 due to emphasis from policy makers on economic growth. With the prevalence of environmental problems, many economists embraced the notion that, "If environmental sustainability must coexist for economic sustainability, then the overall system must permit identification of an equilibrium between the environment and the economy." As such, economic policy makers began to incorporate the functions of the natural environment or natural capital particularly as a sink for wastes and for the provision of raw materials and amenities.

Debate continues among economists as to how to account for natural capital, specifically whether resources can be replaced through knowledge and technology, or whether the environment is a closed system that cannot be replenished and is finite. Economic models influence environmental resource management, in that management policies reflect beliefs about natural capital scarcity. For someone who believes natural capital is infinite and easily substituted, environmental management is irrelevant to the economy. For example, economic paradigms based on neoclassical models of closed economic systems are primarily concerned with resource scarcity, and thus prescribe legalizing the environment as an economic externality for an environmental resource management strategy. This approach has often been termed 'Command-and-control'. Colby has identified trends in the development of economic paradigms, among them, a shift towards more ecological economics since the 1990s.

Ecology

A diagram showing the juvenile fish bypass system, which allows young salmon and steelhead to safely pass the Rocky Reach Hydro Project in Washington.

"The pairing of significant uncertainty about the behaviour and response of ecological systems with urgent calls for near-term action constitutes a difficult reality, and a common lament" for many environmental resource managers. Scientific analysis of the environment deals with several dimensions of ecological uncertainty. These include: structural uncertainty resulting from the misidentification, or lack of information pertaining to the relationships between ecological variables; parameter uncertainty referring to "uncertainty associated with parameter values that are not known precisely but can be assessed and reported in terms of the likelihood of experiencing a defined range of outcomes"; and stochastic uncertainty stemming from chance or unrelated factors. Adaptive management is considered a useful framework for dealing with situations of high levels of uncertainty though it is not without its detractors.

Fencing separates big game from vehicles along the Quebec Autoroute 73 in Canada.

A common scientific concept and impetus behind environmental resource management is carrying capacity. Simply put, carrying capacity refers to the maximum number of organisms a particular resource can sustain. The concept of carrying capacity, whilst understood by many cultures over history, has its roots in Malthusian theory. An example is visible in the EU Water Framework Directive. However, "it is argued that Western scientific knowledge is often insufficient to deal with the full complexity of the interplay of variables in environmental resource management. These concerns have been recently addressed by a shift in environmental resource management approaches to incorporate different knowledge systems including traditional knowledge, reflected in approaches such as adaptive co-management community-based natural resource management and transitions management. among others.

Sustainability

Sustainability in environmental resource management involves managing economic, social, and ecological systems both within and outside an organizational entity so it can sustain itself and the system it exists in. In context, sustainability implies that rather than competing for endless growth on a finite planet, development improves quality of life without necessarily consuming more resources. Sustainably managing environmental resources requires organizational change that instills sustainability values that portrays these values outwardly from all levels and reinforces them to surrounding stakeholders. The end result should be a symbiotic relationship between the sustaining organization, community, and environment.

Many drivers compel environmental resource management to take sustainability issues into account. Today's economic paradigms do not protect the natural environment, yet they deepen human dependency on biodiversity and ecosystem services. Ecologically, massive environmental degradation and climate change threaten the stability of ecological systems that humanity depends on. Socially, an increasing gap between rich and poor and the global North-South divide denies many access to basic human needs, rights, and education, leading to further environmental destruction. The planet's unstable condition is caused by many anthropogenic sources. As an exceptionally powerful contributing factor to social and environmental change, the modern organisation has the potential to apply environmental resource management with sustainability principals to achieve highly effective outcomes. To achieve sustainable development with environmental resource management an organisation should work within sustainability principles, including social and environmental accountability, long-term planning; a strong, shared vision; a holistic focus; devolved and consensus decision making; broad stakeholder engagement and justice; transparency measures; trust; and flexibility.

Current Paradigm Shifts

To adjust to today's environment of quick social and ecological changes, some organizations have begun to experiment with new tools and concepts. Those that are more traditional and stick to hierarchical decision making have difficulty dealing with the demand for lateral decision making that supports effective participation. Whether it be a matter of ethics or just strategic advantage organizations are internalizing sustainability principles. Some of the world's largest and most profitable corporations are shifting to sustainable environmental resource management: Ford, Toyota, BMW, Honda, Shell, Du Port, Sta toil, Swiss Re, Hewlett-Packard, and Unilever, among others. An extensive study by the Boston Consulting Group reaching 1,560 business leaders from diverse regions, job positions, expertise in sustainability, industries, and sizes of organizations, revealed the many benefits of sustainable practice as well as its viability.

It is important to note that though sustainability of environmental resource management has improved, corporate sustainability, for one, has yet to reach the majority of global companies operating in the markets. The three major barriers to preventing organizations to shift towards sustainable practice with environmental resource management are not understanding what sustainability is; having difficulty modeling an economically viable case for the switch; and having a flawed execution plan, or a lack thereof. Therefore, the most important part of shifting an organization to adopt sustainability in environmental resource management would be to create a shared vision and understanding of what sustainability is for that particular organization, and to clarify the business case.

Stakeholders

Public Sector

The public sector comprises the general government sector plus all public corporations including the central bank. In environmental resource management the public sector is responsible for administering natural resource management and implementing environmental protection legislation. The traditional role of the public sector in environmental resource management is to provide professional judgement through skilled technicians on behalf of the public. With the increase of intractable environmental problems, the public sector has been led to examine alternative paradigms for managing environmental resources. This has resulted in the public sector working collaboratively with other sectors (including other governments, private and civil) to encourage sustainable natural resource management behaviours.

A conservation project in North Carolina involving the search for bog turtles was conducted by United States Fish and Wildlife Service and the North Carolina Wildlife Resources Commission and its volunteers.

Private Sector

The private sector comprises private corporations and non-profit institutions serving households. The private sector's traditional role in environmental resource management is that of the recovery of natural resources. Such private sector recovery groups include mining (minerals and petroleum), forestry and fishery organisations. Environmental resource management undertaken by the private sectors varies dependent upon the resource type, that being renewable or non-renewable and private and common resources (also see Tragedy of the Commons). Environmental managers from the private sector also need skills to manage collaboration within a dynamic social and political environment.

Civil Society

Civil society comprises associations in which societies voluntarily organise themselves into and which represent a wide range of interests and ties. These can include community-based organisations, indigenous peoples organisations and non-government organisations (NGO). Functioning through strong public pressure, civil society can exercise their legal rights against the implementation of resource management plans, particularly land management plans. The aim of civil society in environmental resource management is to be included in the decision-making process by means of public participation. Public participation can be an effective strategy to invoke a sense of social responsibility of natural resources.

Sustainability and Environmental Management

At the global scale sustainability and environmental management involves managing the oceans, freshwater systems, land and atmosphere, according to sustainability principles.

Land use change is fundamental to the operations of the biosphere because alterations in the relative proportions of land dedicated to urbanisation, agriculture, forest, woodland, grassland and pasture have a marked effect on the global water, carbon and nitrogen biogeochemical cycles. Management of the Earth's atmosphere involves assessment of all aspects of the carbon cycle to identify opportunities to address human-induced climate change and this has become a major focus of scientific research because of the potential catastrophic effects on biodiversity and human communities. Ocean circulation patterns have a strong influence on climate and weather and, in turn, the food supply of both humans and other organisms.

Atmosphere

In March 2009, at a meeting of the Copenhagen Climate Council 2,500 climate experts from 80 countries issued a keynote statement that there is now no excuse for failing to act on global warming and that without strong carbon reduction targets "abrupt or irreversible" shifts in climate may occur that "will be very difficult for contemporary societies to cope with". Management of the global atmosphere now involves assessment of all aspects of the carbon cycle to identify opportunities to address human-induced climate change and this has become a major focus of scientific research because of the potential catastrophic effects on biodiversity and human communities.

Other human impacts on the atmosphere include the air pollution in cities, the pollutants including toxic chemicals like nitrogen oxides, sulphur oxides, volatile organic compounds and airborne particulate matter that produce photochemical smog and acid rain, and the chlorofluorocarbons that degrade the ozone layer. Anthropogenic particulates such as sulfate aerosols in the atmosphere reduce the direct irradiance and reflectance (albedo) of the Earth's surface. Known as global dimming the decrease is estimated to have been about 4% between 1960 and 1990 although the trend has subsequently reversed. Global dimming may have disturbed the global water cycle by reducing evaporation and rainfall in some areas. It also creates a cooling effect and this may have partially masked the effect of greenhouse gases on global warming.

Oceans

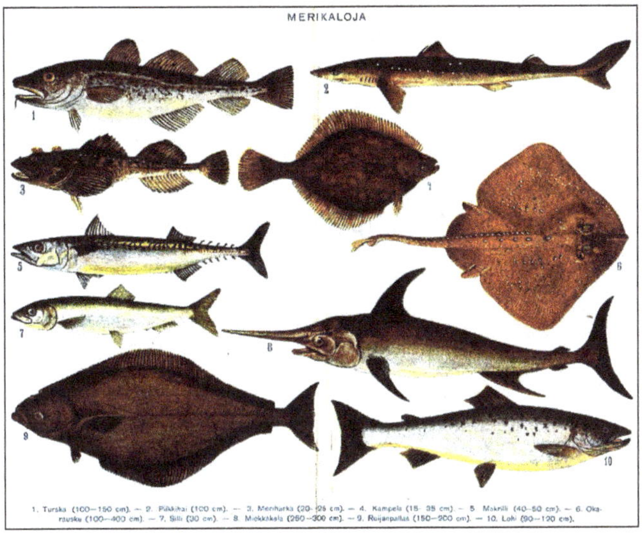

A selection of the world's saltwater fish.

Ocean circulation patterns have a strong influence on climate and weather and, in turn, the food supply of both humans and other organisms. Scientists have warned of the possibility, under the influence of climate change, of a sudden alteration in circulation patterns of ocean currents that could drastically alter the climate in some regions of the globe. Major human environmental impacts occur in the more habitable regions of the ocean fringes the estuaries, coastline and bays. Ten per cent of the world's population about 600 million people live in low-lying areas vulnerable to sea level rise. Trends of concern that require management include: over-fishing (beyond sustainable levels); coral bleaching due to ocean warming and ocean acidification due to increasing levels of dissolved carbon dioxide; and sea level rise due to climate change. Because of their vastness oceans also act as a convenient dumping ground for human waste. Remedial strategies include: more careful waste management, statutory control of overfishing by adoption of sustainable fishing practices and the use of environmentally sensitive and sustainable aquaculture and fish farming, reduction of fossil fuel emissions and restoration of coastal and other marine habitat.

Freshwater

Water covers 71% of the Earth's surface. Of this, 97.5% is the salty water of the oceans and only 2.5% freshwater, most of which is locked up in the Antarctic ice sheet. The remaining freshwater is found in lakes, rivers, wetlands, the soil, aquifers and atmosphere. All life depends on the solar-powered global water cycle, the evaporation from oceans and land to form water vapour that later condenses from clouds as rain, which then becomes the renewable part of the freshwater supply. Awareness of the global importance of preserving water for ecosystem services has only recently emerged as, during the 20th century, more than half the world's wetlands have been lost along with their valuable environmental services. Biodiversity-rich freshwater ecosystems are currently declining faster than marine or land ecosystems making them the world's most vulnerable habitats. Increasing urbanization pollutes clean water supplies and much of the world still does not have access to clean, safe water. In the industrial world demand management has slowed absolute usage rates but increasingly water is being transported over vast distances from water-rich natural areas to population-dense urban areas and energy-hungry desalination is becoming more widely used. Greater emphasis is now being placed on the improved management of blue (harvestable) and green (soil water available for plant use) water, and this applies at all scales of water management.

Land

Loss of biodiversity stems largely from the habitat loss and fragmentation produced by the human appropriation of land for development, forestry and agriculture as natural capital is progressively converted to man-made capital. Land-use change is fundamental to the operations of the biosphere because alterations in the relative proportions of land dedicated to urbanisation, agriculture, forest, woodland, grassland and pasture have a marked effect on the global water, carbon and nitrogen biogeochemical cycles and this can negatively impact both natural and human systems. At the local human scale major sustainability benefits accrue from the pursuit of green cities and sustainable parks and gardens.

Forests

Beech forest – Grib Skov, Denmark.

Since the Neolithic Revolution, human use has reduced the world's forest cover by about 47%. Present-day forests occupy about a quarter of the world's ice-free land with about half of these occurring in the tropics In temperate and boreal regions forest area is gradually increasing (with the exception of Siberia), but deforestation in the tropics is of major concern.

Forests moderate the local climate and the global water cycle through their light reflectance (albedo) and evapotranspiration. They also conserve biodiversity, protect water quality, preserve soil and soil quality, provide fuel and pharmaceuticals, and purify the air. These free ecosystem services are not given a market value under most current economic systems, and so forest conservation has little appeal when compared with the economic benefits of logging and clearance which, through soil degradation and organic decomposition returns carbon dioxide to the atmosphere. The United Nations Food and Agriculture Organization (FAO) estimates that about 90% of the carbon stored in land vegetation is locked up in trees and that they sequester about 50% more carbon than is present in the atmosphere. Changes in land use currently contribute about 20% of total global carbon emissions (heavily logged Indonesia and Brazil are a major source of emissions). Climate change can be mitigated by sequestering carbon in reafforestation schemes, plantations and timber products. Also wood biomass can be utilized as a renewable carbon-neutral fuel. The FAO has suggested that, over the period 2005–2050, effective use of tree planting could absorb about 10–20% of man-made emissions so monitoring the condition of the world's forests must be part of a global strategy to mitigate emissions and protect ecosystem services. However, climate change may pre-empt this FAO scenario as a study by the International Union of Forest Research Organizations in 2009 concluded that the stress of a 2.5C (4.5F) temperature rise above pre-industrial levels could result in the release of vast amounts of carbon so the potential of forests to act as carbon "sinks" is "at risk of being lost entirely".

Cultivated Land

A rice paddy. Rice, wheat, corn and potatoes make up more than half the world's food supply.

Feeding more than six billion human bodies takes a heavy toll on the Earth's resources. This begins with the appropriation of about 38% of the Earth's land surface and about 20% of its net primary productivity. Added to this are the resource-hungry activities of industrial agribusiness everything from the crop need for irrigation water, synthetic fertilizers and pesticides to the resource costs of food packaging, transport (now a major part of global trade) and retail. Food is essential to life. But the list of environmental costs of food production is a long one: topsoil depletion, erosion and conversion to desert from constant tillage of annual crops; overgrazing; salinization; sodification; waterlogging; high levels of fossil fuel use; reliance on inorganic fertilisers and synthetic organic pesticides; reductions in genetic diversity by the mass use of monocultures; water resource depletion; pollution of waterbodies by run-off and groundwater contamination; social problems including the decline of family farms and weakening of rural communities.

All of these environmental problems associated with industrial agriculture and agribusiness are now being addressed through such movements as sustainable agriculture, organic farming and more sustainable business practices.

Extinctions

The extinct dodo (Raphus cucullatus).

Although biodiversity loss can be monitored simply as loss of species, effective conservation demands the protection of species within their natural habitats and ecosystems. Following human migration and population growth, species extinctions have progressively increased to a rate unprecedented since the Cretaceous–Paleogene extinction event. Known as the Holocene extinction event this current human-induced extinction of species ranks as one of the world's six mass extinction events. Some scientific estimates indicate that up to half of presently existing species may become extinct by 2100. Current extinction rates are 100 to 1000 times their prehuman levels with more than

10% birds and mammals threatened, about 8% of plants, 5% of fish and more than 20% of freshwater species.

The 2008, IUCN Red List warns that long-term droughts and extreme weather put additional stress on key habitats and, for example, lists 1,226 bird species as threatened with extinction, which is one-in-eight of all bird species. The Red List Index also identifies 44 tree species in Central Asia as under threat of extinction due to over-exploitation and human development and threatening the region's forests which are home to more than 300 wild ancestors of modern domesticated fruit and nut cultivars.

Biological Invasions

In many parts of the industrial world, land clearing for agriculture has diminished and here the greatest threat to biodiversity, after climate change, has become the destructive effect of invasive species. Increasingly efficient global transport has facilitated the spread of organisms across the planet. The potential danger of this aspect of globalization is starkly illustrated through the spread of human diseases like HIV AIDS, mad cow disease, bird flu and swine flu, but invasive plants and animals are also having a devastating impact on native biodiversity. Non-indigenous organisms can quickly occupy disturbed land and natural areas where, in the absence of their natural predators, they are able to thrive. At the global scale this issue is being addressed through the Global Invasive Species Information Network but there is improved international biosecurity legislation to minimise the transmission of pathogens and invasive organisms. Also, through CITES legislation there is control the trade in rare and threatened species. Increasingly at the local level public awareness programs are alerting communities, gardeners, the nursery industry, collectors, and the pet and aquarium industries, to the harmful effects of potentially invasive species.

Kudzu (Pueraria lobata) infesting trees in Atlanta, Georgia, USA.

Resistance to Change

The environmental sustainability problem has proven difficult to solve. The modern environmental movement has attempted to solve the problem in a large variety of ways.

But little progress has been made, as shown by severe ecological footprint overshoot and lack of sufficient progress on the climate change problem. Something within the human system in preventing change to a sustainable mode of behavior. That system trait is systemic change resistance. Change resistance is also known as organizational resistance, barriers to change, or policy resistance.

Conservation Biology

Efforts are made to preserve the natural characteristics of Hopetoun Falls, Australia, without affecting visitors access.

Conservation biology is the management of nature and of Earth's biodiversity with the aim of protecting species, their habitats, and ecosystems from excessive rates of extinction and the erosion of biotic interactions. It is an interdisciplinary subject drawing on natural and social sciences, and the practice of natural resource management.

The conservation ethic is based on the findings of conservation biology.

The rapid decline of established biological systems around the world means that conservation biology is often referred to as a "Discipline with a deadline". Conservation biology is tied closely to ecology in researching the population ecology (dispersal, migration, demographics, effective population size, inbreeding depression, and minimum population viability) of rare or endangered species. Conservation biology is concerned with phenomena that affect the maintenance, loss, and restoration of biodiversity and the science of sustaining evolutionary processes that engender genetic, population, species, and ecosystem diversity. The concern stems from estimates suggesting that up to 50% of all species on the planet will disappear within the next 50 years, which has contributed to poverty, starvation, and will reset the course of evolution on this planet.

Conservation biologists research and educate on the trends and process of biodiversity loss, species extinctions, and the negative effect these are having on our capabilities to sustain the well-being of human society. Conservation biologists work in the field and office, in government, universities, non-profit organizations and industry. The topics of their research are diverse, because this is an interdisciplinary network with professional alliances in the biological as well as social sciences. Those dedicated to the cause and profession advocate for a global response to the current biodiversity crisis based on morals, ethics, and scientific reason. Organizations and citizens are responding to the biodiversity crisis through conservation action plans that direct research, monitoring, and education programs that engage concerns at local through global scales.

Concepts and Foundations

Measuring Extinction Rates

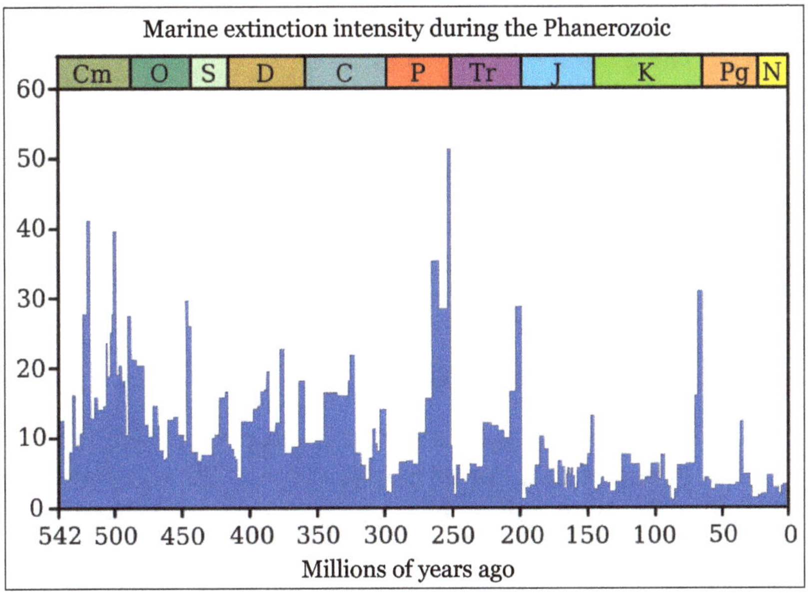

Extinction rates are measured in a variety of ways. Conservation biologists measure and apply statistical measures of fossil records, rates of habitat loss, and a multitude of other variables such as loss of biodiversity as a function of the rate of habitat loss and site occupancy to obtain such estimates. The Theory of Island Biogeography is possibly the most significant contribution toward the scientific understanding of both the process and how to measure the rate of species extinction. The current background extinction rate is estimated to be one species every few years.

The measure of ongoing species loss is made more complex by the fact that most of the Earth's species have not been described or evaluated. Estimates vary greatly on how many species actually exist (estimated range: 3,600,000-111,700,000) to how many have received a species binomial (estimated range: 1.5-8 million). Less than 1% of all species that have been described beyond simply noting its existence. From these figures, the IUCN reports that 23% of vertebrates, 5% of invertebrates and 70% of plants that have been evaluated are designated as endangered or threatened. Better knowledge is being constructed by The Plant List for actual numbers of species.

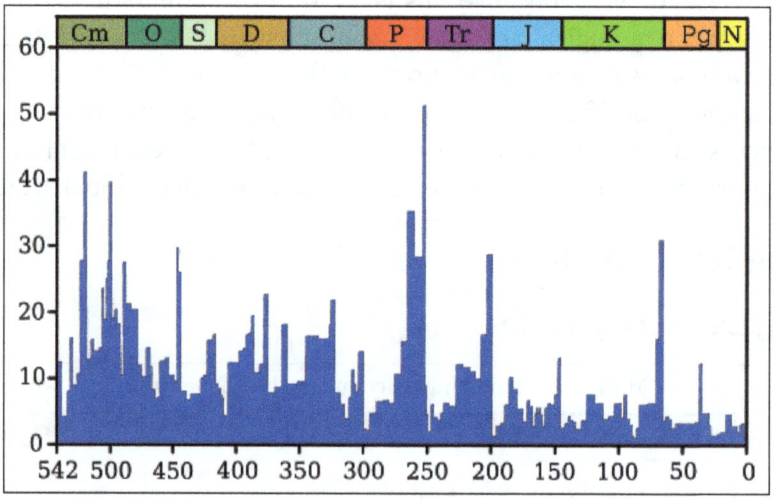

The blue graph shows the apparent percentage (not the absolute number) of marine animal genera becoming extinct during any given time interval. It does not represent all marine species, just those that are readily fossilized. The labels of the traditional "Big Five" extinction events and the more recently recognised.

Systematic Conservation Planning

Systematic conservation planning is an effective way to seek and identify efficient and effective types of reserve design to capture or sustain the highest priority biodiversity values and to work with communities in support of local ecosystems. Margules and Pressey identify six interlinked stages in the systematic planning approach:

- Compile data on the biodiversity of the planning region.
- Identify conservation goals for the planning region.
- Review existing conservation areas.
- Select additional conservation areas.
- Implement conservation actions.
- Maintain the required values of conservation areas.

Conservation biologists regularly prepare detailed conservation plans for grant proposals or to effectively coordinate their plan of action and to identify best management practices (e.g.). Systematic strategies generally employ the services of Geographic Information Systems to assist in the decision making process.

Conservation Physiology: A Mechanistic Approach to Conservation

Conservation physiology was defined by Steven J. Cooke and colleagues as: 'An integrative scientific discipline applying physiological concepts, tools, and knowledge to characterizing biological diversity and its ecological implications; understanding and predicting how organisms, populations, and ecosystems respond to environmental change and stressors; and solving conservation problems across the broad range of taxa (i.e. including microbes, plants, and animals). Physiology is considered in the broadest possible terms to include functional and mechanistic responses at all scales, and conservation includes the development and refinement of strategies to rebuild populations, restore ecosystems, inform conservation policy, generate decision-support tools, and manage natural resources.' Conservation physiology is particularly relevant to practitioners in that it has the potential to generate cause-and-effect relationships and reveal the factors that contribute to population declines.

Approaches

Conservation may be classified as either in-situ conservation, which is protecting an endangered species in its natural habitat, or ex-situ conservation, which occurs outside the natural habitat. In-situ conservation involves protecting or restoring the habitat. Ex-situ conservation, on the other hand, involves protection outside of an organism's natural habitat, such as on reservations or in gene banks, in circumstances where viable populations may not be present in the natural habitat.

Also, non-interference may be used, which is termed a preservationist method. Preservationists advocate for giving areas of nature and species a protected existence that halts interference from the humans. In this regard, conservationists differ from preservationists in the social dimension, as conservation biology engages society and seeks equitable solutions for both society and ecosystems. Some preservationists emphasize the potential of biodiversity in a world without humans.

Ethics and Values

Conservation biologists are interdisciplinary researchers that practice ethics in the biological and social sciences. Chan states that conservationists must advocate for biodiversity and can do so in a scientifically ethical manner by not promoting simultaneous advocacy against other competing values.

A conservationist may be inspired by the resource conservation ethic, which seeks to identify what measures will deliver "the greatest good for the greatest number of people for the longest time." In contrast, some conservation biologists argue that nature has an intrinsic value that is independent of anthropocentric usefulness or utilitarianism. Intrinsic value advocates that a gene, or species, be valued because they have a utility for the ecosystems they sustain. Aldo Leopold was a classical thinker and writer on such conservation ethics whose philosophy, ethics and writings are still valued and revisited by modern conservation biologists.

Conservation Priorities

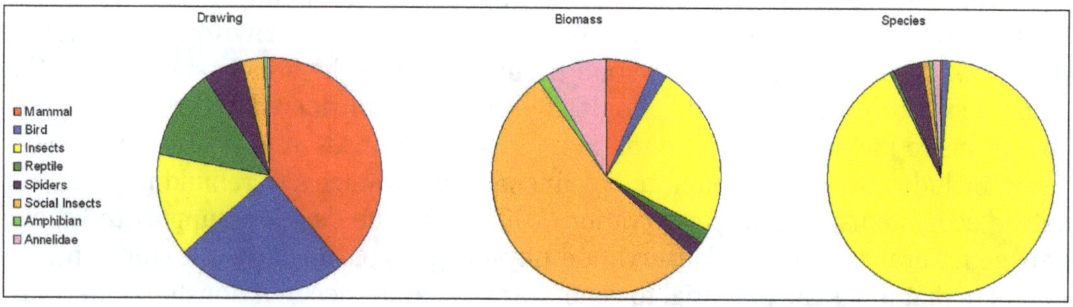

A pie chart image showing the relative biomass representation in a rain forest through a summary of children's perceptions from drawings and artwork (left), through a scientific estimate of actual biomass (middle), and by a measure of biodiversity (right). Notice that the biomass of social insects (middle) far outweighs the number of species (right).

The International Union for the Conservation of Nature (IUCN) has organized a global assortment of scientists and research stations across the planet to monitor the changing state of nature in an effort to tackle the extinction crisis. The IUCN provides annual updates on the status of species conservation through its Red List. The IUCN Red List serves as an international conservation tool to identify those species most in need of conservation attention and by providing a global index on the status of biodiversity. More than the dramatic rates of species loss, however, conservation scientists note that the sixth mass extinction is a biodiversity crisis requiring far more action than a priority focus on rare, endemic or endangered species. Concerns for biodiversity loss covers a broader conservation mandate that looks at ecological processes, such as migration, and a holistic examination of biodiversity at levels beyond the species, including genetic, population and ecosystem diversity. Extensive, systematic, and rapid rates of biodiversity loss threatens the sustained well-being of humanity by limiting supply of ecosystem services that are otherwise regenerated by the complex and evolving holistic network of genetic and ecosystem diversity. While the conservation status of species is employed extensively in conservation management, some scientists highlight that it is the common species that are the primary source of exploitation and habitat alteration by humanity. Moreover, common species are often undervalued despite their role as the primary source of ecosystem services.

While most in the community of conservation science "stress the importance" of sustaining biodiversity, there is debate on how to prioritize genes, species, or ecosystems, which are all components of biodiversity (e.g. Bowen, 1999). While the predominant approach to date has been to focus efforts on endangered species by conserving biodiversity hotspots, some scientists (e.g) and conservation organizations, such as the Nature Conservancy, argue that it is more cost-effective, logical, and socially relevant to invest in biodiversity coldspots. The costs of discovering, naming, and mapping out the distribution of every species, they argue, is an ill-advised conservation venture. They reason it is better to understand the significance of the ecological roles of species.

Biodiversity hotspots and coldspots are a way of recognizing that the spatial concentration of genes, species, and ecosystems is not uniformly distributed on the Earth's surface. For example, "44% of all species of vascular plants and 35% of all species in four vertebrate groups are confined to 25 hotspots comprising only 1.4% of the land surface of the Earth."

Those arguing in favor of setting priorities for coldspots point out that there are other measures to consider beyond biodiversity. They point out that emphasizing hotspots downplays the importance of the social and ecological connections to vast areas of the Earth's ecosystems where biomass, not biodiversity, reigns supreme. It is estimated that 36% of the Earth's surface, encompassing 38.9% of the worlds vertebrates, lacks the endemic species to qualify as biodiversity hotspot. Moreover, measures show that maximizing protections for biodiversity does not capture ecosystem services any better than targeting randomly chosen regions. Population level biodiversity (mostly in coldspots) are disappearing at a rate that is ten times that at the species level. The level of importance in addressing biomass versus endemism as a concern for conservation biology is highlighted in literature measuring the level of threat to global ecosystem carbon stocks that do not necessarily reside in areas of endemism. A hotspot priority approach would not invest so heavily in places such as steppes, the Serengeti, the Arctic, or taiga. These areas contribute a great abundance of population (not species) level biodiversity and ecosystem services, including cultural value and planetary nutrient cycling.

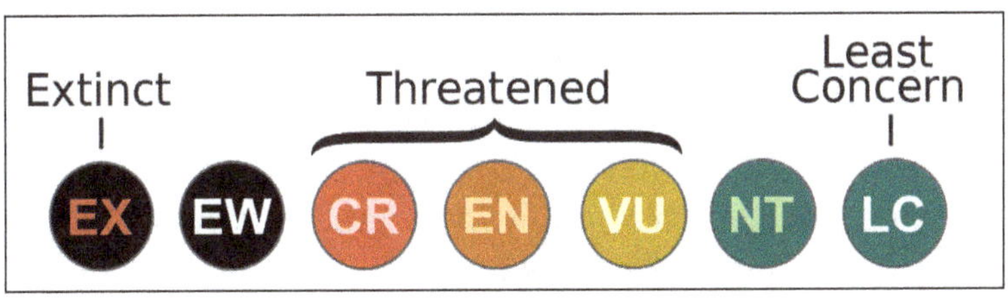

Summary of 2006 IUCN Red List categories.

Those in favor of the hotspot approach point out that species are irreplaceable components of the global ecosystem, they are concentrated in places that are most threatened, and should therefore receive maximal strategic protections. Species richness and genetic biodiversity contributes to and engenders ecosystem stability, ecosystem processes, evolutionary adaptability, and biomass. Both sides agree, however, that conserving biodiversity is necessary to reduce the extinction rate and identify an inherent value in nature; the debate hinges on how to prioritize limited conservation resources in the most cost-effective way.

Economic Values and Natural Capital

Tadrart Acacus desert in western Libya, part of the Sahara.

Conservation biologists have started to collaborate with leading global economists to determine how to measure the wealth and services of nature and to make these values apparent in global market transactions. This system of accounting is called natural capital and would, for example, register the value of an ecosystem before it is cleared to make way for development. The WWF publishes its Living Planet Report and provides a global index of biodiversity by monitoring approximately 5,000 populations in 1,686 species of vertebrate (mammals, birds, fish, reptiles, and amphibians) and report on the trends in much the same way that the stock market is tracked.

This method of measuring the global economic benefit of nature has been endorsed by the G8+5 leaders and the European Commission. Nature sustains many ecosystem services that benefit humanity. Many of the Earth's ecosystem services are public goods without a market and therefore no price or value. When the stock market registers a financial crisis, traders on Wall Street are not in the business of trading stocks for much of the planet's living natural capital stored in ecosystems. There is no natural stock market with investment portfolios into sea horses, amphibians, insects, and other creatures that provide a sustainable supply of ecosystem services that are valuable to society. The ecological footprint of society has exceeded the bio-regenerative capacity limits of the planet's ecosystems by about 30 percent, which is the same percentage of vertebrate populations that have registered decline from 1970 through 2005.

The inherent natural economy plays an essential role in sustaining humanity, including the regulation of global atmospheric chemistry, pollinating crops, pest control, cycling soil nutrients, purifying our water supply, supplying medicines and health benefits, and unquantifiable quality of life improvements. There is a relationship, a correlation, between markets and natural capital, and social income inequity and biodiversity loss. This means that there are greater rates of biodiversity loss in places where the inequity of wealth is greatest.

Although a direct market comparison of natural capital is likely insufficient in terms of human value, one measure of ecosystem services suggests the contribution amounts to trillions of dollars yearly. For example, one segment of North American forests has been assigned an annual value of 250 billion dollars; as another example, honey-bee pollination is estimated to provide between 10 and 18 billion dollars of value yearly. The value of ecosystem services on one New Zealand island has been imputed to be as great as the GDP of that region. This planetary wealth is being lost at an incredible rate as the demands of human society is exceeding the bio-regenerative capacity of the Earth. While biodiversity and ecosystems are resilient, the danger of losing them is that humans cannot recreate many ecosystem functions through technological innovation.

Strategic Species Concepts

Keystone Species

Some species, called a keystone species form a central supporting hub unique to their ecosystem. The loss of such a species results in a collapse in ecosystem function, as well as the loss of coexisting species. Keystone species are usually predators due to their ability to control the population of prey in their ecosystem. The importance of a keystone species was shown by the extinction of the Steller's sea cow (Hydrodamalis gigas) through its interaction with sea otters, sea urchins, and kelp. Kelp beds grow and form nurseries in shallow waters to shelter creatures that support the food chain. Sea urchins feed on kelp, while sea otters feed on sea urchins. With the rapid decline of sea otters due to overhunting, sea urchin populations grazed unrestricted on the kelp beds and the ecosystem collapsed. Left unchecked, the urchins destroyed the shallow water kelp communities that supported the Steller's sea cow's diet and hastened their demise. The sea otter was thought to be a keystone species because the coexistence of many ecological associates in the kelp beds relied upon otters for their survival. However this was later questioned by Turvey and Risley, who showed that hunting alone would have driven the Steller's sea cow extinct.

Indicator Species

An indicator species has a narrow set of ecological requirements, therefore they become useful targets for observing the health of an ecosystem. Some animals, such as amphibians with their semi-permeable skin and linkages to wetlands, have an acute sensitivity to environmental harm and thus may serve as a miner's canary. Indicator species are monitored in an effort to capture environmental degradation through pollution or some other link to proximate human activities. Monitoring an indicator species is a measure to determine if there is a significant environmental impact that can serve to advise or modify practice, such as through different forest silviculture treatments and management scenarios, or to measure the degree of harm that a pesticide may impart on the health of an ecosystem.

Government regulators, consultants, or NGOs regularly monitor indicator species, however, there are limitations coupled with many practical considerations that must be followed for the approach to be effective. It is generally recommended that multiple indicators (genes, populations, species, communities, and landscape) be monitored for effective conservation measurement that prevents harm to the complex, and often unpredictable, response from ecosystem dynamics.

Umbrella and Flagship Species

An example of an umbrella species is the monarch butterfly, because of its lengthy migrations and aesthetic value. The monarch migrates across North America, covering multiple ecosystems and so requires a large area to exist. Any protections afforded to the monarch butterfly will at the same time umbrella many other species and habitats. An umbrella species is often used as flagship species, which are species, such as the giant panda, the blue whale, the tiger, the mountain gorilla and the monarch butterfly, that capture the public's attention and attract support for conservation measures. Paradoxically, however, conservation bias towards flagship species sometimes threatens other species of chief concern.

Context and Trends

Conservation biologists study trends and process from the paleontological past to the ecological present as they gain an understanding of the context related to species extinction. It is generally accepted that there have been five major global mass extinctions that register in Earth's history. These include: the Ordovician (440 mya), Devonian (370 mya), Permian–Triassic (245 mya), Triassic–Jurassic (200 mya), and Cretaceous–Paleogene extinction event (66 mya) extinction spasms. Within the last 10,000 years, human influence over the Earth's ecosystems has been so extensive that scientists have difficulty estimating the number of species lost; that is to say the rates of deforestation, reef destruction, wetland draining and other human acts are proceeding much faster than human assessment of species. The latest *Living Planet Report* by the World Wide Fund for Nature estimates that we have exceeded the bio-regenerative capacity of the planet, requiring 1.6 Earths to support the demands placed on our natural resources.

Holocene Extinction

Conservation biologists are dealing with and have published evidence from all corners of the planet indicating that humanity may be causing the sixth and fastest planetary extinction event. It has been suggested that an unprecedented number of species is becoming extinct in what is known as the Holocene extinction event. The global extinction rate may be approximately 1,000 times higher than the natural background extinction rate. It is estimated that two-thirds of all mammal genera and one-half of all mammal species weighing at least 44 kilograms (97 lb) have gone extinct in the last 50,000 years. The Global Amphibian Assessment reports that amphibians are declining on a global scale faster than any other vertebrate group, with over 32% of all surviving species being threatened with extinction. The surviving populations are in continual decline in 43% of those that are threatened. Since the mid-1980s the actual rates of extinction have exceeded 211 times rates measured from the fossil record. However, "The current amphibian extinction rate may range from 25,039 to 45,474 times the background extinction rate for amphibians." The global extinction trend occurs in every major vertebrate group that is being monitored. For example, 23% of all mammals and 12% of all birds are Red Listed by the International Union for Conservation of Nature (IUCN), meaning they too are threatened with extinction. Even though extinction is natural, the decline in species is happening at such an incredible rate that evolution can simply not match, therefore, leading to the greatest continual mass extinction on Earth. Humans have dominated the planet and our high consumption of resources, along with the pollution generated is affecting the environments in which other species live. There are a wide variety of species that humans are working to protect such as the Hawaiian Crow and the Whooping Crane of Texas. People can also take action on preserving species by advocating and voting for global and national policies that improve climate, under the concepts of climate mitigation and climate restoration. The Earth's oceans demand particular attention as climate change continues to alter pH levels, making it uninhabitable for organisms with shells which dissolve as a result.

An art scape image showing the relative importance of animals in a rain forest through a summary of (a) child's perception compared with (b) a scientific estimate of the importance. The size of the animal represents its importance. The child's mental image places importance on big cats, birds, butterflies, and then reptiles versus the actual dominance of social insects (such as ants).

Status of Oceans and Reefs

Global assessments of coral reefs of the world continue to report drastic and rapid rates of decline. By 2000, 27% of the world's coral reef ecosystems had effectively collapsed. The largest period of decline occurred in a dramatic "bleaching" event in 1998, where approximately 16% of all the coral reefs in the world disappeared in less than a year. Coral bleaching is caused by a mixture of environmental stresses, including increases in ocean temperatures and acidity, causing both the release of symbiotic algae and death of corals. Decline and extinction risk in coral reef biodiversity has risen dramatically in the past ten years. The loss of coral reefs, which are predicted to go extinct in the next century, threatens the balance of global biodiversity, will have huge economic impacts, and endangers food security for hundreds of millions of people. Conservation biology plays an important role in international agreements covering the world's oceans (and other issues pertaining to biodiversity).

The oceans are threatened by acidification due to an increase in CO_2 levels. This is a most serious threat to societies relying heavily upon oceanic natural resources. A concern is that the majority of all marine species will not be able to evolve or acclimate in response to the changes in the ocean chemistry.

The prospects of averting mass extinction seems unlikely when "90% of all of the large (average approximately ≥50 kg), open ocean tuna, billfishes, and sharks in the ocean" are reportedly gone. Given the scientific review of current trends, the ocean is predicted to have few surviving multi-cellular organisms with only microbes left to dominate marine ecosystems.

Groups other than Vertebrates

Serious concerns also being raised about taxonomic groups that do not receive the same degree of social attention or attract funds as the vertebrates. These include fungal (including lichen-forming species), invertebrate (particularly insect) and plant communities where the vast majority of biodiversity is represented. Conservation of fungi and conservation of insects, in particular, are both of pivotal importance for conservation biology. As mycorrhizal symbionts, and as decomposers and recyclers, fungi are essential for sustainability of forests. The value of insects in the biosphere is enormous because they outnumber all other living groups in measure of species richness. The greatest bulk of biomass on land is found in plants, which is sustained by insect relations. This great ecological value of insects is countered by a society that often reacts negatively toward these aesthetically 'unpleasant' creatures.

One area of concern in the insect world that has caught the public eye is the mysterious case of missing honey bees (Apis mellifera). Honey bees provide an indispensable ecological services through their acts of pollination supporting a huge variety of agriculture crops. The use of honey and wax have become vastly used throughout the world. The sudden disappearance of bees leaving empty hives or colony collapse disorder (CCD) is not uncommon. However, in 16-month period from 2006 through 2007, 29% of 577 beekeepers across the United States reported CCD losses in up to 76% of their colonies. This sudden demographic loss in bee numbers is placing a strain on the agricultural sector. The cause behind the massive declines is puzzling scientists. Pests, pesticides, and global warming are all being considered as possible causes.

Another highlight that links conservation biology to insects, forests, and climate change is the mountain pine beetle (Dendroctonus ponderosae) epidemic of British Columbia, Canada, which has infested 470,000 km^2 (180,000 sq mi) of forested land since 1999. An action plan has been prepared by the Government of British Columbia to address this problem.

Conservation Biology of Parasites

A large proportion of parasite species are threatened by extinction. A few of them are being eradicated as pests of humans or domestic animals, however, most of them are harmless. Threats include the decline or fragmentation of host populations, or the extinction of host species.

Threats to Biodiversity

Today, many threats to Biodiversity exist. An acronym that can be used to express the top threats of present-day H.I.P.P.O stands for Habitat Loss, Invasive Species, Pollution, Human Population, and Overharvesting. The primary threats to biodiversity are habitat destruction (such as deforestation, agricultural expansion, urban development), and overexploitation (such as wildlife trade). Habitat fragmentation also poses challenges, because the global network of protected areas only covers 11.5% of the Earth's surface. A significant consequence of fragmentation and lack of linked protected areas is the reduction of animal migration on a global scale. Considering that billions of tonnes of biomass are responsible for nutrient cycling across the earth, the reduction of migration is a serious matter for conservation biology.

However, human activities need not necessarily cause irreparable harm to the biosphere. With conservation management and planning for biodiversity at all levels, from genes to ecosystems, there are examples where humans mutually coexist in a sustainable way with nature. Even with the current threats to biodiversity there are ways we can improve the current condition and start anew.

Many of the threats to biodiversity, including disease and climate change, are reaching inside borders of protected areas, leaving them 'not-so protected' (e.g. Yellowstone National Park). Climate change, for example, is often cited as a serious threat in this regard, because there is a feedback loop between species extinction and the release of carbon dioxide into the atmosphere. Ecosystems store and cycle large amounts of carbon which regulates global conditions. In present day, there have been major climate shifts with temperature changes making survival of some species difficult. The effects of global warming add a catastrophic threat toward a mass extinction of global biological diversity. Conservationists have claimed that not all the species can be saved, and they have to decide which their efforts should be used to protect. This concept is known as the Conservation Triage. The extinction threat is estimated to range from 15 to 37 percent of all species by 2050, or 50 percent of all species over the next 50 years. The current extinction rate is 100-100,000 times more rapid today than the last several billion years.

Soil Conservation

Erosion barriers on disturbed slope, Marin County, California.

Contour plowing in Pennsylvania.

Soil conservation is the prevention of soil loss from erosion or prevention of reduced fertility caused by over usage, acidification, salinization or other chemical soil contamination.

Slash-and-burn and other unsustainable methods of subsistence farming are practiced in some lesser developed areas. A sequel to the deforestation is typically large scale erosion, loss of soil nutrients and sometimes total desertification. Techniques for improved soil conservation include crop rotation, cover crops, conservation tillage and planted windbreaks, affect both erosion and fertility. When plants die, they decay and become part of the soil. Code 330 defines standard methods recommended by the U.S. Natural Resources Conservation Service. Farmers have practiced soil conservation for millennia. In Europe, policies such as the Common Agricultural Policy are targeting the application of best management practices such as reduced tillage, winter cover crops, plant residues and grass margins in order to better address the soil conservation. Political and economic action is further required to solve the erosion problem. A simple governance hurdle concerns how we value the land and this can be changed by cultural adaptation.

Contour Ploughing

Contour ploughing orients furrows following the contour lines of the farmed area. Furrows move left and right to maintain a constant altitude, which reduces runoff. Contour ploughing was practiced by the ancient Phoenicians, and is effective for slopes between two and ten percent. Contour ploughing can increase crop yields from 10 to 50 percent, partially as a result of greater soil retention.

Terrace Farming

Terracing is the practice of creating nearly level areas in a hillside area. The terraces form a series of steps, each at a higher level than the previous. Terraces are protected from erosion by other soil barriers. Terraced farming is more common on small farms and in underdeveloped countries, since mechanized equipment is difficult to deploy in this setting.

Keyline Design

Keyline design is an enhancement of contour farming, where the total watershed properties are taken into account in forming the contour lines.

Perimeter Runoff Control

Tree, shrubs and ground-cover are effective perimeter treatment for soil erosion prevention, by impeding surface flows. A special form of this perimeter or inter-row treatment is the use of a "grass way" that both channels and dissipates runoff through surface friction, impeding surface runoff and encouraging infiltration of the slowed surface water.

Windbreaks

Windbreaks are sufficiently dense rows of trees at the windward exposure of an agricultural field subject to wind erosion. Evergreen species provide year-round protection;

however, as long as foliage is present in the seasons of bare soil surfaces, the effect of deciduous trees may be adequate.

Cover Crops/Crop Rotation

Cover crops such as legumes plant, white turnip, radishes and other species are rotated with cash crops to blanket the soil year-round and act as green manure that replenishes nitrogen and other critical nutrients. Cover crops also help suppress weeds.

Soil-conservation Farming

Soil-conservation farming involves no-till farming, "green manures" and other soil-enhancing practices. Such farming methods attempt to mimic the biology of barren lands. They can revive damaged soil, minimize erosion, encourage plant growth, eliminate the use of nitrogen fertilizer or fungicide, produce above-average yields and protect crops during droughts or flooding. The result is less labor and lower costs that increase farmers' profits. No-till farming and cover crops act as sinks for nitrogen and other nutrients. This increases the amount of soil organic matter.

Repeated plowing/tilling degrades soil, killing its beneficial fungi and earthworms. Once damaged, soil may take multiple seasons to fully recover, even in optimal circumstances.

Critics argue that no-till and related methods are impractical and too expensive for many growers, partly because it requires new equipment. They cite advantages for conventional tilling depending on the geography, crops and soil conditions. Some farmers claimed that no-till complicates weed control, delays planting and that post-harvest residues, especially for corn, are hard to manage.

Salinity Management

Salinity in soil is caused by irrigating with salty water. Water then evaporates from the soil leaving the salt behind. Salt breaks down the soil structure, causing infertility and reduced growth.

The ions responsible for salination are: sodium (Na^+), potassium (K^+), calcium (Ca^{2+}), magnesium (Mg^{2+}) and chlorine (Cl^-). Salinity is estimated to affect about one third of the earth's arable land. Soil salinity adversely affects crop metabolism and erosion usually follows.

Salinity occurs on drylands from overirrigation and in areas with shallow saline water tables. Over-irrigation deposits salts in upper soil layers as a byproduct of soil infiltration; irrigation merely increases the rate of salt deposition. The best-known case of shallow saline water table capillary action occurred in Egypt after the 1970 construction of the Aswan Dam. The change in the groundwater level led to high salt concentrations in the water table. The continuous high level of the water table led to soil salination.

Salt deposits on the former bed of the Aral Sea.

Use of humic acids may prevent excess salination, especially given excessive irrigation. Humic acids can fix both anions and cations and eliminate them from root zones.

Planting species that can tolerate saline conditions can be used to lower water tables and thus reduce the rate of capillary and evaporative enrichment of surface salts. Salt-tolerant plants include saltbush, a plant found in much of North America and in the Mediterranean regions of Europe.

Soil Organisms

When worms excrete feces in the form of casts, a balanced selection of minerals and plant nutrients is made into a form accessible for root uptake. Earthworm casts are five times richer in available nitrogen, seven times richer in available phosphates and eleven times richer in available potash than the surrounding upper 150 millimetres (5.9 in) of soil. The weight of casts produced may be greater than 4.5 kg per worm per year. By burrowing, the earthworm improves soil porosity, creating channels that enhance the processes of aeration and drainage.

Yellow fungus, a mushroom that assists in organic decay.

Other important soil organisms include nematodes, mycorrhiza and bacteria. Degraded soil requires synthetic fertilizer to produce high yields. Lacking structure increases erosion and carries nitrogen and other pollutants into rivers and streams. Each one percent increase in soil organic matter helps soil hold 20,000 gallons more water per acre.

Mineralization

To allow plants full realization of their phytonutrient potential, active mineralization of the soil is sometimes undertaken. This can involve adding crushed rock or chemical soil supplements. In either case the purpose is to combat mineral depletion. A broad range of minerals can be used, including common substances such as phosphorus and more exotic substances such as zinc and selenium. Extensive research examines the phase transitions of minerals in soil with aqueous contact.

Flooding can bring significant sediments to an alluvial plain. While this effect may not be desirable if floods endanger life or if the sediment originates from productive land, this process of addition to a floodplain is a natural process that can rejuvenate soil chemistry through mineralization.

Watershed Management

Watershed is defined as a geohydrological unit draining to a common point by a system of drains. All lands on earth are part of one watershed or other. Watershed is thus the land and water area, which contributes runoff to a common point. A watershed is an area of land and water bounded by a drainage divide within which the surface runoff collects and flows out of the watershed through a single outlet into a lager river or lake.

The watershed area for a small stream may be a few hectares, while for a large river the watershed may be many square kilometres. For proper planning and execution, the size of a watershed area should ideally be 1,000 to 2,500 hectares. If the area is within this range, it will be possible to prepare a well balanced plan and to implement it in a period of 2 to 3 years.

Objectives of Watershed Management

Watershed management is an attempt to halt land degradation and a holistic process for getting maximum production out of land. Watershed management implies rational utilisation of land and water resources for optimum and sustained production, with the minimum of hazard to natural resources. With a bit of acumen and proper judgement, the rain water can be made to move down the slopes slowly, ensuring optimum infiltration and percolation, automatically solving the problem of soil erosion. Reducing the impact of the rain on the soil, checking its speed at various intervals, taking up all operations on the contour and diverting the excess to prevent the, pressure, are some of the procedures one could adopt; a process that starts at the highest point of the watershed and ends right down in the lower parts.

Steps in Watershed Planning

- Preparation of base maps for carrying out surveys.
- Reconnaissance survey of the watershed for overall development.
- Assessing rainfall characteristics.
- Preparation of soil maps and classification of lands for different uses according to capability classification for agriculture, forestry, pasture, horticulture, etc.
- Preparation of inventory of existing land uses and farm sizes.
- Appraisal of agricultural production patterns and potentials, present and potential markets and possible group action arrangements.
- Carrying out topographic and hydrologic surveys for engineering works.
- Geo-hydrological survey to delineate areas suitable for groundwater development.
- Formulation of an integrated time-bound plan for land and moisture conservation, ground water. recharge, development of productive afforestation, agriculture production, grasslands and horticulture.
- Assigning of priorities for implementation of the project.
- Assessing social costs and benefits.

Land Capability Classification

Watershed management must take into account land capability classification, since a typical watershed may include lands suitable for cropping, trees, pastures, etc. The land classification is as follows:

Class I

This type of land is nearly level, with minimum erosion, the soils are deep, well drained, with adequate water holding capacity. These soils need ordinary management practices to maintain productivity. Such practices may include the use of one or more of the following fertilisers, lime cover and green manure crops, conservation of crop residues aAd crop rotation. They are fit for agriculture, pasture, forests, etc.

Class II

These soils have some limitations. They are subject to slight erosion, as they are slightly sloping lands, with less soil depth, occasional drainage problems that can be easily corrected, and they are sometimes saline. They may need one or more of the following practices terracing, strip-cropping, contour cultivation, excess water drainage, crop rotation, use of green manure crops, stubble mulching, the use of fertilisers, manure and lime.

These soils may be used for growing cultivated crops, raising pastures, forests, etc.

Class III

The factors that limit plant growth are more severe than in class II. These factors could be moderately sloping land, land susceptible to water or wind erosion, land more susceptible to drainage problems, land with water-logging, shallow soil depth, low moisture holding capacity and moderate salinity.

Class IV

The limitations to plant growth is more severe here than in Class III soils. Very careful management is required and the conservation practices are more difficult to apply and maintain. Some of the factors affecting these lands are steep slopes, severe susceptibility to water and wind erosion, shallow soils, low moisture holding capacity, severe water-logging and severe salinity. These soils can be used for crops, pastures, forests and wildlife food, etc.

Class V

Cultivation is not feasible because of one or more limitations, such as over flow, stoniness, wetness or severe climatic conditions. Examples of this class are soils of low lands subject to frequent over flows which prevent the normal production of cultivated crops, or rocky or stony terrain. These soils are suitable for raising cultivated crops but are not suitable for perennial vegetation.

Class VI

These soils have very severe limitations that make them generally unsuitable for cultivation and limit their use largely to pastures or forests or wildlife food and cover. Some of the limitations which cannot be corrected are: steep slope, very severe erosion hazard, shallow rooting zone, excessive wetness, low moisture capacity and severe salinity.

Class VII

The physical condition of the soil is such that it is not practical to adopt pasture improvements and water control practices.

Class VIII

Bad lands, rocky out crops, sandy beaches, marshes, deserts, mine tailings and other nearly barren lands may be included in this class. It may be necessary to give protection and management for plant growth to soils and land forms in class VIII in order to protect other more valuable soils, to control water, or for wildlife or for aesthetic reasons.

Water Management Activities

The best opportunities for moisture conservation lie in the control of the portion of precipitation which is normal runoff. Retaining surface water on the land surface for a longer time, increases the amount of water entering the soil. Effective methods are the following:

- Contour bunding / graded bunding.
- Check dams and gully control structures.
- Land levelling / land smoothening.
- Bench terracing.
- Farm ponds.
- Percolation ponds.
- Waterways.
- Diversion drains.

Water conservation starts in the topmost rocky areas where the runoff water needs to be drained out by forming diversion drains. This protects the steep slopes below which are suited only for growing fodder. While prevention of grazing in these areas ensures optimum regeneration, luxuriant fodder cut and fed to the animals ensures better nutrition and health.

References

- Reed, M.S.; Graves, A.; Dandy, N.; Posthumus, H.; Hubacek, K.; Morris, J.; Prell, C.; Quinn, C.H.; Stinger, L.C. (2009). "Who's in and why? A typology of stakeholder analysis methods for natural resource management". Journal of Environmental Management. 90 (5): 1933–1949. doi:10.1016/j.jenvman.2009.01.001. PMID 19231064
- Pahl-Wost, C. (May 2007). "The implications of complexity for integrated resource management". Environmental Modelling and Software. 22 (5): 561–9. CiteSeerX 10.1.1.196.1136. doi:10.1016/j.envsoft.2005.12.024
- Watershed-management, energy-environment-know-your-environment-water; vikaspedia.in, Retrieved 19 February, 2020
- Ruckelshaus, Mary; Klinger, Terrie; Knowlton, Nancy; Demaster, Douglas P. (2008). "Marine Ecosystem-based Management in Practice: Scientific and Governance Challenges". BioScience. 58: 53–63. doi:10.1641/B580110
- Meffe, Gary K.; Martha J. Groom (2006). Principles of conservation biology (3rd ed.). Sunderland, Mass: Sinauer Associates. ISBN 978-0-87893-518-5
- Panagos, Panos; Borrelli, Pasquale; Meusburger, Katrin; Alewell, Christine; Lugato, Emanuele; Montanarella, Luca (2015). "Estimating the soil erosion cover-management factor at the European scale". Land Use Policy. 48: 38–50. doi:10.1016/j.landusepol.2015.05.021

4

Natural Resources and their Management

Natural resources are categorized into renewable and non-renewable resources. Renewable resources include sunlight, air, water, etc. whereas non-renewable resources include coal, petroleum and fossil fuels. Management of these resources for maintaining the quality of life is referred to as natural resource management. This chapter discusses natural resources and their management in detail.

Natural Resource

Natural resources refer to the things that exist freely in nature for human use and don't necessarily need the action of mankind for their generation or production. The key aspect of natural resources is that they dictate the survival of humans and other life forms on earth. These resources include land, rocks, forests (vegetation), water (ocean, lakes, streams, seas, and rivers), fossil fuel, animals (fish, wild life, and domesticated animals), minerals, sunlight and air.

Some examples of natural resources are: air which provides wind energy, Coal which act as an input for electricity, forests which provide paper, wood and various medicines, Water which is used for drinking and production of hydroelectric energy, sunlight that is used for drying clothes, photosynthesis and solar energy.

They are known as Natural Resources because they provide for the basis of life on earth. It is from the natural resources that humans obtain and produce the components and materials found within our environments. Every artificial product is made from the natural resources. The materials may be used as they occur naturally or may be transformed in other forms.

However, most natural resources are prone to depletion and degradation which has brought about worldwide concerns for their sustainable usage and management. Still, there are other very few resources that are regarded to be inexhaustible such as sunlight and geothermal energy.

Renewable Resources

A renewable resource is a natural resource which will replenish to replace the portion depleted by usage and consumption, either through natural reproduction or other recurring processes in a finite amount of time in a human time scale. Renewable resources are a part of Earth's natural environment and the largest components of its ecosphere. A positive life cycle assessment is a key indicator of a resource's sustainability.

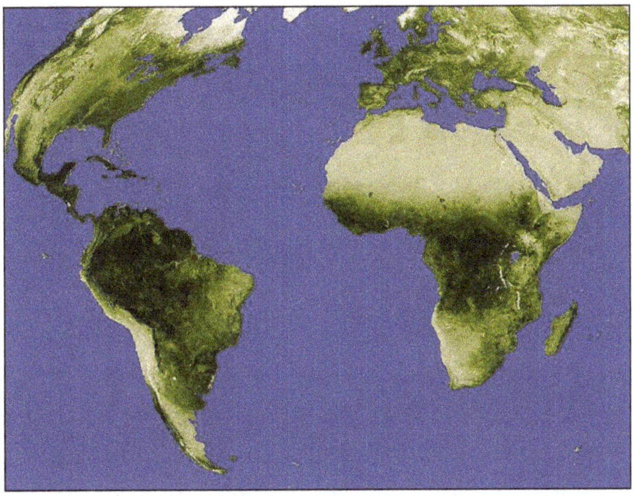

Global vegetation.

Oceans often act as renewable resources.

Definitions of renewable resources may also include agricultural production, as in sustainable agriculture and to an extent water resources. In 1962, Paul Alfred Weiss defined renewable resources as: "The total range of living organisms providing man with life, fibres, etc". Another type of renewable resources is renewable energy resources. Common sources of renewable energy include solar, geothermal and wind power, which are all categorised as renewable resources.

Sawmill near Fügen, Zillertal, Austria.

Air, Food and Water

Water Resources

Water can be considered a renewable material when carefully controlled usage, treatment, and release are followed. If not, it would become a non-renewable resource at that location. For example, as groundwater is usually removed from an aquifer at a rate much greater than its very slow natural recharge, it is a considered non-renewable resource. Removal of water from the pore spaces in aquifers may cause permanent compaction (subsidence) that cannot be renewed. 97.5% of the water on the Earth is salt water, and 3% is fresh water; slightly over two thirds of this is frozen in glaciers and polar ice caps. The remaining unfrozen freshwater is found mainly as groundwater, with only a small fraction (0.008%) present above ground or in the air.

Water pollution is one of the main concerns regarding water resources. It is estimated that 22% of worldwide water is used in industry. Major industrial users include hydroelectric dams, thermoelectric power plants (which use water for cooling), ore and oil refineries (which use water in chemical processes) and manufacturing plants (which use water as a solvent),it is also used for dumping garbage.

Desalination of seawater is considered a renewable source of water, although reducing its dependence on fossil fuel energy is needed for it to be fully renewable.

Panorama of a natural wetland (Sinclair Wetlands, New Zealand).

Non Agricultural Food

Alaska wild "berries" from the Innoko National Wildlife Refuge - Renewable Resources.

Food is any substance consumed to provide nutritional support for the body. Most food has its origin in renewable resources. Food is obtained directly from plants and animals.

Hunting may not be the first source of meat in the modernised world, but it is still an important and essential source for many rural and remote groups. It is also the sole source of feeding for wild carnivores.

Sustainable Agriculture

The phrase sustainable agriculture was coined by Australian agricultural scientist Gordon McClymont. It has been defined as "an integrated system of plant and animal production practices having a site-specific application that will last over the long term". Expansion of agricultural land reduces biodiversity and contributes to deforestation. The Food and Agriculture Organization of the United Nations estimates that in coming decades, cropland will continue to be lost to industrial and urban development, along with reclamation of wetlands, and conversion of forest to cultivation, resulting in the loss of biodiversity and increased soil erosion.

Polyculture practices in Andhra Pradesh.

Although air and sunlight are available everywhere on Earth, crops also depend on soil nutrients and the availability of water. Monoculture is a method of growing only one crop at a time in a given field, which can damage land and cause it to become either unusable or suffer from reduced yields. Monoculture can also cause the build-up of pathogens and pests that target one specific species. The Great Irish Famine (1845–1849) is a well-known example of the dangers of monoculture.

Crop rotation and long-term crop rotations confer the replenishment of nitrogen through the use of green manure in sequence with cereals and other crops, and can improve soil structure and fertility by alternating deep-rooted and shallow-rooted plants. Other methods to combat lost soil nutrients are returning to natural cycles that annually flood cultivated lands (returning lost nutrients indefinitely) such as the Flooding of the Nile, the long-term use of biochar, and use of crop and livestock landraces that are adapted to less than ideal conditions such as pests, drought, or lack of nutrients.

Agricultural practices are one of the single greatest contributor to the global increase in soil erosion rates. It is estimated that "more than a thousand million tonnes of southern Africa's soil are eroded every year. Experts predict that crop yields will be halved within thirty to fifty years if erosion continues at present rates." The Dust Bowl phenomenon in the 1930s was caused by severe drought combined with farming methods that did not include crop rotation, fallow fields, cover crops, soil terracing and wind-breaking trees to prevent wind erosion.

The tillage of agricultural lands is one of the primary contributing factors to erosion, due to mechanised agricultural equipment that allows for deep plowing, which severely increases the amount of soil that is available for transport by water erosion. The phenomenon called *peak soil* describes how large-scale factory farming techniques are affecting humanity's ability to grow food in the future. Without efforts to improve soil management practices, the availability of arable soil may become increasingly problematic.

Illegal slash and burn practice in Madagascar.

Methods to combat erosion include no-till farming, using a keyline design, growing wind breaks to hold the soil, and widespread use of compost. Fertilizers and pesticides can also have an effect of soil erosion, which can contribute to soil salinity and prevent other species from growing. Phosphate is a primary component in the chemical fertiliser applied most commonly in modern agricultural production. However, scientists estimate that rock phosphate reserves will be depleted in 50–100 years and that Peak Phosphate will occur in about 2030.

Industrial processing and logistics also have an effect on agriculture's sustainability. The way and locations crops are sold requires energy for transportation, as well as the energy cost for materials, labour, and transport. Food sold at a local location, such a farmers' market, have reduced energy overheads.

Air

Air is a renewable resource. All living organisms need oxygen, nitrogen (directly or indirectly), carbon (directly or indirectly) and many other gases in small quantities for their survival.

Non-food Resources

Douglas fir forest created in 1850, Meymac (Corrèze), France.

An important renewable resource is wood provided by means of forestry, which has been used for construction, housing and firewood since ancient times. Plants provide the main sources for renewable resources, the main distinction is between energy crops and non-food crops. A large variety of lubricants, industrially used vegetable oils, textiles and fibre made e.g. of cotton, copra or hemp, paper derived from wood, rags or grasses, bioplastic are based on plant renewable resources. A large variety of chemical based products like latex, ethanol, resin, sugar and starch can be provided with plant renewables. Animal based renewables include fur, leather, technical fat and lubricants and further derived products, as e.g. animal glue, tendons, casings or in historical times ambra and baleen provided by whaling.

With regard to pharmacy ingredients and legal and illegal drugs, plants are important sources, however e.g. venom of snakes, frogs and insects has been a valuable renewable source of pharmacological ingredients. Before GMO production set in, insulin and important hormones were based on animal sources. Feathers, an important byproduct of poultry farming for food, is still being used as filler and as base for keratin in general. Same applies for the chitin produced in farming Crustaceans which may be used as base of chitosan. The most important part of the human body used for non-medical purposes is human hair as for artificial hair integrations, which is being traded worldwide.

Renewables used for Self Sufficiency

In vitro-culture of Vitis (grapevine), Geisenheim Grape Breeding Institute.

The success of the German chemical industry till World War I was based on the replacement of colonial products. The predecessors of IG Farben dominated the world market for synthetic dyes at the beginning of the 20th century and had an important role in artificial pharmaceuticals, photographic film, agricultural chemicals and electrochemicals.

However, the former Plant breeding research institutes took a different approach. After the loss of the German colonial empire, important players in the field as Erwin Baur and Konrad Meyer switched to using local crops as base for economic autarky. Meyer as a key agricultural scientist and spatial planner of the Nazi era managed and lead

Deutsche Forschungsgemeinschaft resources and focused about a third of the complete research grants in Nazi Germany on agricultural and genetic research and especially on resources needed in case of a further German war effort. A wide array of agrarian research institutes still existing today and having importance in the field was founded or enlarged in the time.

There were some major failures as trying to e.g. grow frost resistant olive species, but some success in the case of hemp, flax, rapeseed, which are still of current importance. During World War 2, German scientists tried to use Russian Taraxacum (dandelion) species to manufacture natural rubber. Rubber dandelions are still of interest, as scientists in the Fraunhofer Institute for Molecular Biology and Applied Ecology (IME) announced 2013 to have developed a cultivar that is suitable for commercial production of natural rubber.

Examples of Industrial Use

Biorenewable Chemicals

Biorenewable chemicals are chemicals created by biological organisms that provide feedstocks for the chemical industry. Biorenewable chemicals can provide solar-energy-powered substitutes for the petroleum-based carbon feedstocks that currently supply the chemical industry. The tremendous diversity of enzymes in biological organisms, and the potential for synthetic biology to alter these enzymes to create yet new chemical functionalities, can drive the chemical industry. A major platform for creation of new chemicals is the polyketide biosynthetic pathway, which generates chemicals containing repeated alkyl chain units with potential for a wide variety of functional groups at the different carbon atoms.

Bioplastics

A packaging blister made from cellulose acetate, a bioplastic.

Bioplastics are a form of plastics derived from renewable biomass sources, such as vegetable fats and oils, lignin, corn starch, pea starch or microbiota. The most common form of bioplastic is thermoplastic starch. Other forms include Cellulose bioplastics, biopolyester, Polylactic acid, and bio-derived polyethylene.

The production and use of bioplastics is generally regarded as a more sustainable activity when compared to plastic production from petroleum (petroplastic); however, manufacturing of bioplastic materials is often still reliant upon petroleum as an energy and materials source. Because of the fragmentation in the market and ambiguous definitions it is difficult to describe the total market size for bioplastics, but the global production capacity is estimated at 327,000 tonnes. In contrast, global consumption of all flexible packaging is estimated at around 12.3 million tonnes.

Bioasphalt

Bioasphalt is an asphalt alternative made from non-petroleum based renewable resources. Manufacturing sources of bioasphalt include sugar, molasses and rice, corn and potato starches, and vegetable oil based waste. Asphalt made with vegetable oil based binders was patented by Colas SA in France in 2004.

Renewable Energy

Renewable energy refers to the provision of energy via renewable resources which are naturally replenished fast enough as being used. It includes e.g. sunlight, wind, biomass, rain, tides, waves and geothermal heat. Renewable energy may replace or enhance fossil energy supply various distinct areas: electricity generation, hot water/ space heating, motor fuels, and rural (off-grid) energy services.

Biomass

A sugarcane plantation in Brazil (State of São Paulo). Cane is used for biomass energy.

Biomass is referring to biological material from living, or recently living organisms, most often referring to plants or plant-derived materials.

Sustainable harvesting and use of renewable resources (i.e., maintaining a positive renewal rate) can reduce air pollution, soil contamination, habitat destruction and land degradation. Biomass energy is derived from six distinct energy sources: garbage, wood, plants, waste, landfill gases, and alcohol fuels. Historically, humans have harnessed biomass-derived energy since the advent of burning wood to make fire, and wood remains the largest biomass energy source today.

However, low tech use of biomass, which still amounts for more than 10% of world energy needs may induce indoor air pollution in developing nations and results in between 1.5 million and 2 million deaths in 2000.

The biomass used for electricity generation varies by region. Forest by-products, such as wood residues, are common in the United States. Agricultural waste is common in Mauritius (sugar cane residue) and Southeast Asia (rice husks). Animal husbandry residues, such as poultry litter, are common in the UK. The biomass power generating industry in the United States, which consists of approximately 11,000 MW of summer operating capacity actively supplying power to the grid, produces about 1.4 percent of the U.S. electricity supply.

Biofuel

Brazil has bioethanol made from sugarcane available throughout the country. Shown a typical Petrobras gas station at São Paulo with dual fuel service, marked A for alcohol (ethanol) and G for gasoline.

A biofuel is a type of fuel whose energy is derived from biological carbon fixation. Biofuels include fuels derived from biomass conversion, as well as solid biomass, liquid fuels and various biogases.

Bioethanol is an alcohol made by fermentation, mostly from carbohydrates produced in sugar or starch crops such as corn, sugarcane or switchgrass.

Biodiesel is made from vegetable oils and animal fats. Biodiesel is produced from oils or fats using transesterification and is the most common biofuel in Europe.

Biogas is methane produced by the process of anaerobic digestion of organic material by anaerobes., etc. is also a renewable source of energy.

Biogas

Biogas typically refers to a mixture of gases produced by the breakdown of organic matter in the absence of oxygen. Biogas is produced by anaerobic digestion with anaerobic bacteria or fermentation of biodegradable materials such as manure, sewage, municipal waste, green waste, plant material, and crops. It is primarily methane (CH_4) and carbon dioxide (CO_2) and may have small amounts of hydrogen sulphide (H_2S), moisture and siloxanes.

Natural Fibre

Natural fibres are a class of hair-like materials that are continuous filaments or are in discrete elongated pieces, similar to pieces of thread. They can be used as a component of composite materials. They can also be matted into sheets to make products such as paper or felt. Fibres are of two types: natural fibre which consists of animal and plant fibres, and man made fibre which consists of synthetic fibres and regenerated fibres.

Threats to Renewable Resources

Renewable resources are endangered by non-regulated industrial developments and growth. They must be carefully managed to avoid exceeding the natural world's capacity to replenish them. A life cycle assessment provides a systematic means of evaluating renewability. This is a matter of sustainability in the natural environment.

Overfishing

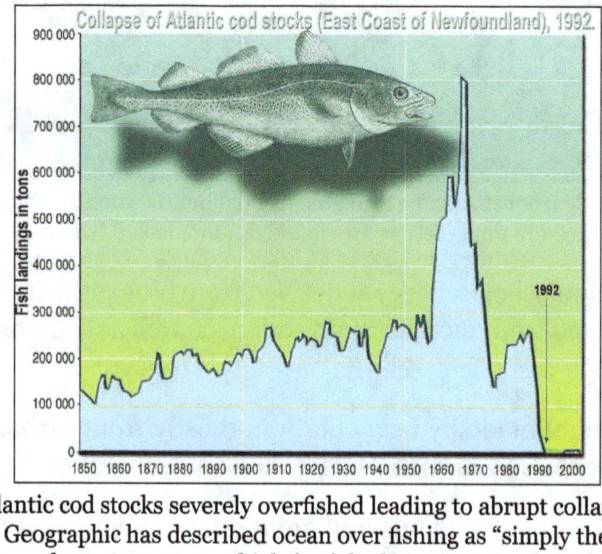

Atlantic cod stocks severely overfished leading to abrupt collapse.
National Geographic has described ocean over fishing as "simply the taking of wildlife from the sea at rates too high for fished species to replace themselves."

Tuna meat is driving overfishing as to endanger some species like the bluefin tuna. The European Community and other organisations are trying to regulate fishery as to protect species and to prevent their extinctions.

Examples of overfishing exist in areas such as the North Sea of Europe, the Grand Banks of North America and the East China Sea of Asia.

The decline of penguin population is caused in part by overfishing, caused by human competition over the same renewable resources.

Deforestation in Brazil.

Deforestation

Besides their role as a resource for fuel and building material, trees protect the environment by absorbing carbon dioxide and by creating oxygen. The destruction of rain forests is one of the critical causes of climate change. Deforestation causes carbon dioxide to linger in the atmosphere. As carbon dioxide accrues, it produces a layer in the atmosphere that traps radiation from the sun. The radiation converts to heat which causes global warming, which is better known as the greenhouse effect.

Deforestation also affects the water cycle. It reduces the content of water in the soil and groundwater as well as atmospheric moisture. Deforestation reduces soil cohesion, so that erosion, flooding and landslides ensue.

Rain forests house many species and organisms providing people with food and other commodities. In this way biofuels may well be unsustainable if their production contributes to deforestation.

Endangered Species

Over-hunting of American Bison.

Some renewable resources, species and organisms are facing a very high risk of extinction caused by growing human population and over-consumption. It has been estimated that over 40% of all living species on Earth are at risk of going extinct. Many nations have laws to protect hunted species and to restrict the practice of hunting. Other conservation methods include restricting land development or creating preserves. The IUCN Red List of Threatened Species is the best-known worldwide conservation status listing and ranking system. Internationally, 199 countries have signed an accord agreeing to create Biodiversity Action Plans to protect endangered and other threatened species.

Non-renewable Resource

A coal mine in Wyoming, United States. Coal, produced over millions of years, is a finite and non-renewable resource on a human time scale.

A non-renewable resource (also called a finite resource) is a resource of economic value that cannot be readily replaced by natural means at a quick enough pace to keep up with consumption. An example is carbon-based fossil fuel. The original organic matter, with the aid of heat and pressure, becomes a fuel such as oil or gas. Earth minerals and metal ores, fossil fuels (coal, petroleum, natural gas) and groundwater in certain aquifers are all considered non-renewable resources, though individual elements are always conserved (except in nuclear reactions).

Conversely, resources such as timber (when harvested sustainably) and wind (used to power energy conversion systems) are considered renewable resources, largely because their localized replenishment can occur within time frames meaningful to humans as well.

Earth Minerals and Metal Ores

Earth minerals and metal ores are examples of non-renewable resources. The metals themselves are present in vast amounts in Earth's crust, and their extraction by humans only occurs where they are concentrated by natural geological processes (such as heat, pressure, organic activity, weathering and other processes) enough to become economically viable to extract. These processes generally take from tens of thousands to millions of years, through plate tectonics, tectonic subsidence and crustal recycling.

The localized deposits of metal ores near the surface which can be extracted economically by humans are non-renewable in human time-frames. There are certain rare earth minerals and elements that are more scarce and exhaustible than others. These are in high demand in manufacturing, particularly for the electronics industry.

Most metal ores are considered vastly greater in supply to fossil fuels, because metal ores are formed by crustal-scale processes which make up a much larger portion of the Earth's near-surface environment, than those that form fossil fuels which are limited to areas where carbon-based life forms flourish, die, and are quickly buried.

Fossil Fuels

Natural resources such as coal, petroleum (crude oil) and natural gas take thousands of years to form naturally and cannot be replaced as fast as they are being consumed. Eventually it is considered that fossil-based resources will become too costly to harvest and humanity will need to shift its reliance to other sources of energy such as solar or wind power.

An alternative hypothesis is that carbon based fuel is virtually inexhaustible in human terms, if one includes all sources of carbon-based energy such as methane hydrates on the sea floor, which are vastly greater than all other carbon based fossil fuel resources combined. These sources of carbon are also considered non-renewable, although their rate of formation/replenishment on the sea floor is not known. However their extraction at economically viable costs and rates has yet to be determined.

At present, the main energy source used by humans is non-renewable fossil fuels. Since the dawn of internal combustion engine technologies in the 19th century, petroleum and other fossil fuels have remained in continual demand. As a result, conventional infrastructure and transport systems, which are fitted to combustion engines, remain prominent throughout the globe. The continual use of fossil fuels is known to contribute to global warming and cause more severe climate change.

Nuclear Fuels

Rössing uranium mine is the longest-running and one of the largest open pit uranium mines in the world, in 2005 it produced eight percent of global uranium oxide needs(3,711 tons). The most productive mines however are the underground McArthur River uranium mine in Canada which produces 13% of the world's uranium, and the similarly underground poly-metallic Olympic Dam mine in Australia, which despite being largely a copper mine, contains the largest known reserve of uranium ore.

Annual release of "technologically enhanced"/concentrated Naturally occurring radioactive material, uranium and thorium radioisotopes naturally found in coal and concentrated in heavy/bottom coal ash and airborne fly ash. As predicted by ORNL to cumulatively amount to 2.9 million tons over the 1937-2040 period, from the combustion of an estimated 637 billion tons of coal worldwide. This 2.9 million tons of actinide fuel, a resource derived from coal ash, would be classified as low grade uranium ore if it occurred naturally.

In 1987, the World Commission on Environment and Development (WCED) an organization set up by but independent from the United Nations classified fission reactors that produce more fissile nuclear fuel than they consume -i.e. breeder reactors, and when it is developed, fusion power, among conventional renewable energy sources, such as solar and falling water. The American Petroleum Institute likewise does not consider conventional nuclear fission as renewable, but that breeder reactor nuclear power fuel is considered renewable and sustainable, before explaining that radioactive waste from used spent fuel rods remains radioactive, and so has to be very carefully stored for up to a thousand years. With the careful monitoring of radioactive waste products also being required upon the use of other renewable energy sources, such as geothermal energy.

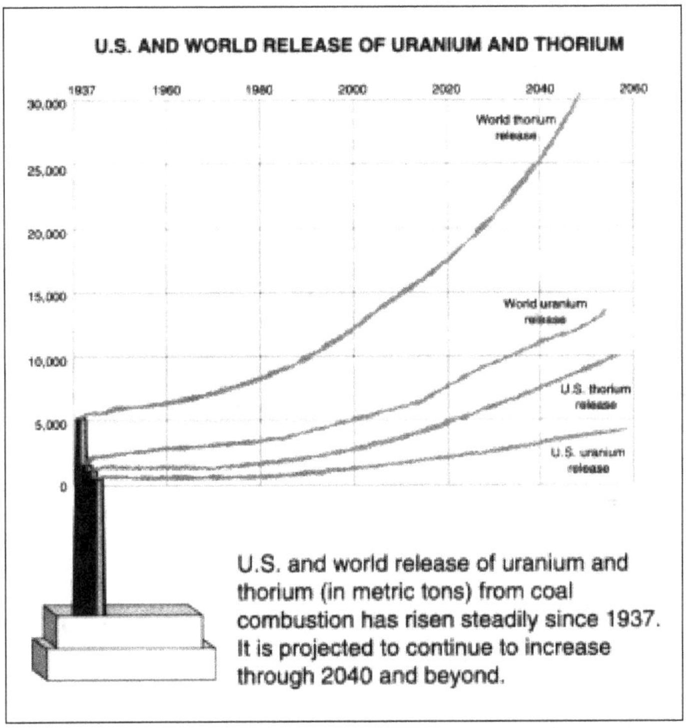

U.S. and world release of uranium and thorium (in metric tons) from coal combustion has risen steadily since 1937. It is projected to continue to increase through 2040 and beyond.

The use of nuclear technology relying on fission requires Naturally occurring radioactive material as fuel. Uranium, the most common fission fuel, and is present in the ground at relatively low concentrations and mined in 19 countries. This mined uranium is used to fuel energy-generating nuclear reactors with fissionable uranium-235 which generates heat that is ultimately used to power turbines to generate electricity.

As of 2013, only a few kilograms (picture available) of uranium have been extracted from the ocean in pilot programs and it is also believed that the uranium extracted on an industrial scale from the seawater would constantly be replenished from uranium leached from the ocean floor, maintaining the seawater concentration at a stable level.

Nuclear power provides about 6% of the world's energy and 13–14% of the world's electricity. Nuclear energy production is associated with potentially dangerous radioactive contamination as it relies upon unstable elements. In particular, nuclear power facilities produce about 200,000 metric tons of low and intermediate level waste (LILW) and 10,000 metric tons of high level waste (HLW) (including spent fuel designated as waste) each year worldwide.

Issues entirely separate from the question of the sustainability of nuclear fuel, relate to the use of nuclear fuel and the high-level radioactive waste the nuclear industry generates that if not properly contained, is highly hazardous to people and wildlife. The United Nations (UNSCEAR) estimated in 2008 that average annual human radiation exposure includes 0.01 millisievert (mSv) from the legacy of past atmospheric nuclear testing plus the Chernobyl disaster and the nuclear fuel cycle, along with 2.0 mSv from natural radioisotopes and 0.4 mSv from cosmic rays; all exposures vary by location. natural uranium in some inefficient reactor nuclear fuel cycles, becomes part of the nuclear waste "once through" stream, and in a similar manner to the scenario were this uranium remained naturally in the ground, this uranium emits various forms of radiation in a decay chain that has a half-life of about 4.5 billion years, the storage of this unused uranium and the accompanying fission reaction products have raised public concerns about risks of leaks and containment, however the knowledge gained from studying the Natural nuclear fission reactor in Oklo Gabon, has informed geologists on the proven processes that kept the waste from this 2 billion year old natural nuclear reactor that operated for hundreds of thousands of years.

Biotic Resources

Biotic resources are also known as the living resources These are the renewable resources and include the plants and animals. The foods, sugars, tea, coffee, medicines are obtained from the plants. The timber is obtained from the tress and is used for the construction, furniture, ships, poles and radio. The wood pulp forms different types of paper. The gums, wool, acetic acid, alcohol, and dry fruits are obtained from the trees. They also provide the shelter and food to the other organisms. They also provide oxygen and affect the weather of a particular area. The plants present on the land and are classified as a forests, croplands, shrubs and grassland. The vegetation affects the soil by preventing soil erosion. The humus is formed by the decayed leaves which provide nutrition to the soil. The spread of deserts is also checked by the plants. In a year the annual production of biomass produced by the marine plants is three hundred twenty five million tones. The total production of biomass is three hundred eighty million tones. The humans mostly depend on the land vegetation for survival. The earth has around 4 million species of plants and a few of them are edible. The conservation of plants must be done. The pollution in the environment, deforestation, overgrazing, and the formation of cropland causes an adverse effect on the growth of plants. Some of the species of plants get endangered. So, there should be a check on the activities of human. The use of marine plants is multipurpose. The humans must explore the plants and which act as a source of food. The role of animals in the biotic environment is very crucial. They act as a food for the man. The food is provided by the prawns, fish, goat, sheep, and pig and honey Bess. Some of them are used as a mode of transport. They include the camel, horse, asses, horses and mules. Some of them are used as a mode of agriculture operations. They include the cattle and camel's.Some of them is used as a scavenger. They include the jackals and pigs. Some of them are used as pearls. They include the mollusks. The animals maintain the ecosystem. The fecal matter of certain animals is used as manure. The deer and elephant provide the musk and ivory. The wax and lac is provided by the honey bees and lac insect. The pollination and dispersal of seeds and fruits is mainly done by the insects. Some animals are used as a laboratory research model like the guinea pigs, rats and rabbits. The diseases of human beings are cured by the secretion of hormones of some mammals. The conservation of animals must be done. The pollution in the environment, deforestation, killing and the poaching of animals causes an adverse effect on the growth of animals. Some of the species of animals get endangered. So, there should be a check on the activities of human. In the coming 25 years around one fourth of the species of animals are going to disappear. The people must be educated regarding the wild life of animals. The micro organisms as the name suggests consists of small organisms which cannot be seen with the help of a naked eye. They are present almost everywhere on the earth. The Anton van Leeuwenhoek was the first one to discover them. It consists of the bacteria, protozoa, viruses and fungi. The fungi involve the yeast and moulds. Bacteria are present everywhere.

The size of bacteria is not static. It is different for every species. Their size ranges from 0.1 micrometer to 750 micrometer. Bacterial cells vary in shape and they have no definite structure. They can be coccus, bacilli, spirillum, stalked in shape. The fungi are organisms without chlorophyll, have heterotrophic mode of nutrition. They are spore forming, avascular and eukaryotic organism. Their cell wall is made up of chitin. They store their food in the form of glycogen. The fungi form a medicine known as penicillin which is used to treat the bacterial infections. The fungi also acts on the dead organisms and act as a saprophyte. They can create a problem also. The protozoa can act as a harmful as well as beneficial organism. The harmful organisms act to pollute water, decrease the soil fertility and cause certain diseases. They also act as a food for the aquatic animals. The oil deposits are found in association with the skeletal deposits of radiolarian and other organisms. They help to locate the fuel. Almost all the viruses are harmful.

Exploitation of Natural Resources

The exploitation of natural resources is the use of natural resources for economic growth, sometimes with a negative connotation of accompanying environmental degradation. It started to emerge on an industrial scale in the 19th century as the extraction and processing of raw materials (such as in mining, steam power, and machinery) developed much further than it had in preindustrial areas. During the 20th century, energy consumption rapidly increased. Today, about 80% of the world's energy consumption is sustained by the extraction of fossil fuels, which consists of oil, coal and gas. Another non-renewable resource that is exploited by humans is subsoil minerals such as precious metals that are mainly used in the production of industrial commodities. Intensive agriculture is an example of a mode of production that hinders many aspects of the natural environment, for example the degradation of forests in a terrestrial ecosystem and water pollution in an aquatic ecosystem. As the world population rises and economic growth occurs, the depletion of natural resources influenced by the unsustainable extraction of raw materials becomes an increasing concern.

Timber.

Reasons for Resources being under Pressure

- Increase in the sophistication of technology enabling natural resources to be extracted quickly and efficiently. E.g., in the past, it could take long hours just to cut down one tree only using saws. Due to increased technology, rates of deforestation have greatly increased.
- The number of humans is increasing. According to the UN, there were 7.6 billion of us in 2017. This number is expected to rise to about 10 billion in 2050 and about 11 billion in 2100.
- Cultures of consumerism. Materialistic views lead to the mining of gold and diamonds to produce jewelry, unnecessary commodities for human life or advancement. Consumerism also leads to extraction of resources for the production of commodities necessary for human life but in amounts excessive of what is needed, because people consume more than is necessary or waste what they have.
- Excessive demand often leads to conflicts due to intense competition. Organizations such as Global Witness and the United Nations have documented the connection.

- Lack of awareness among the population is striking. People are not aware of ways to reduce depletion and exploitation of materials.

Consequences of Exploitation of Resources

Natural resources are not limitless, and the following consequences can arise from the careless and excessive consumption of these resources:

- Deforestation.
- Desertification.
- Extinction of species.
- Forced migration.
- Soil erosion.
- Oil depletion.
- Ozone depletion.
- Greenhouse gas increase.
- Extreme energy.
- Water pollution.
- Natural hazard/Natural disaster.
- Metals and minerals depletion.

Effects on Local Communities

The Global South

Human Resources Macon, Georgia.

When a mining company enters a developing country in the global south to extract raw materials, advocating the advantages of the industry's presence and minimizing the potential negative effects gain cooperation of the local people. Advantageous factors are primarily in economic development so services that the government could not provide such as health centers, police departments and schools can be established. However, with economic development, money becomes a dominant subject of interest. This can bring about major conflicts that a local community in a developing country has never dealt with before. These conflicts emerge by a change to more egocentric views among the locals influenced by consumerist values.

The effects of the exploitation of natural resources in the local community of a developing country are exhibited in the impacts from the Ok Tedi Mine. After BHP, now BHP Billiton, entered into Papua New Guinea to exploit copper and gold, the economy of the indigenous peoples boomed. Although their quality of life has improved, initially disputes were common among the locals in terms of land rights and who should be getting the benefits from the mining project. The consequences of the Ok Tedi environmental disaster illustrate the potential negative effects from the exploitation of natural resources. The resulting mining pollution includes toxic contamination of the natural water supply for communities along the Ok Tedi River, causing widespread killing of aquatic life. When a mining company ends a project after extracting the raw materials from an area of a developing country, the local people are left to manage with the environmental damage done to their community and the long run sustainability of the economic benefits stimulated by the mining company's presence becomes a concern.

Natural Resource Management

Natural resource management are the ways in which societies manage the supply of or access to the natural resources upon which they rely for their survival and development. Insofar as humans are fundamentally dependent on natural resources, ensuring the ongoing access to or a steady provision of natural resources has always been central to the organization of civilizations and, historically, has been organized through a range of schemes varying in degrees of formality and involvement from the central authorities.

A "natural" resource is one that is afforded by nature without human intervention; hence, the fertile lands or the minerals within them, rather than the crop that grows on them, are examples of a country's natural resources. Although what is considered a "resource" (or, for that matter, "natural") has varied over time and from one society to another, resources are, ultimately, riches provided by nature from which can be derived some form of benefit, whether material or immaterial. Under some definitions, only those natural resources that can renew themselves and whose exploitation relies on their regenerative capacities properly necessitate management. For example, oil is not usually considered a subject of natural resource management, whereas forests are. The use of nonrenewable resources is subject to regulation rather than management. The management of renewable natural resources seeks to balance the demands of exploitation with a respect for regenerative capacities.

Forestry.
A logging operation in Scotland.

The emergence of a rational systematic management of natural resources can be traced back to the phase of accelerated industrialization of the late 19th century. In a period of unprecedented industrial growth, the pressures brought to bear on the supply of raw materials and natural resources by an unrelenting demand intensified the need to rationalize their utilization, so as to eliminate increasingly costly waste and to allocate them more efficiently. That coincided with a broader tendency toward rationalization, a general social pattern identified by the sociologist Max Weber that emerged in modern industrial societies in response to the large-scale reorganization of production and whereby goal-oriented rationality was increasingly infused into the organization of social activities. Natural resource management was born at the conjunction of rationalization and its twin process, bureaucratization, which yielded the first bureaucracies to manage nature.

Of course, there were huge variations in both the rates and degrees to which the different states became involved with questions of natural resource management. The French state, for example, took a heavy hand in forestry management as early as the 17th century, when wood became a strategic resource at a time of accelerated mercantilist (export-oriented) growth that relied primarily on maritime transportation namely, wooden ships. Such local variations aside, overall it took a certain kind of state, the modern bureaucratic state, to steer the exploitation of natural resources toward principles of scientific management. In the United States, natural resource management was made a federal matter for the first time under the presidency of Theodore Roosevelt. At that time, principles of scientific management, which combined notions of rational management with in-depth scientific knowledge of the resource itself, were promoted by key figures such as Gifford Pinchot, who took a leading role in the U.S. government's management of forests in the 1890s and served as head of the Forest Service from its creation, in 1905 to 1910. In Europe a similar concern with rational resource exploitation transpired at about the same time. The International Council for the Exploration of the Sea, for example, provided a forum in which northern European countries could share concerns about maritime research and resources. It was effectively one of the first international conferences on a natural resource management question, and there too science was entrenched as a basis for exploitation of the seas, laying the grounds for future arrangement for the management of collective resources.

Encountering Earth's Limits

The 20th century saw natural resource management increasingly projected at a supranational level, where it was also collectivized. A first major impulse toward the internationalization of natural resource management was brought by the post World War II context, with its pervasive spirit of cooperation on the one hand and its specific problems of food shortages on the other. Countries came together to address the issues of damaged capacities and insufficient production in other words, insufficient use of available resources. That context yielded the Food and Agricultural Organization (FAO) in 1945, the International Whaling Commission in 1946, and, much later, the International Fund for Agricultural Development, established in 1977 to tackle problems of agricultural production in developing countries. The problem was seen to lie in the management of the resources rather than in the resources themselves. Therefore, the solution was to develop common solutions to management problems that were widely shared from one country to the next. The problem those organizations attempted to solve, in other words, was how to create international regimes that would disseminate better management solutions and thus enable each country to make better use of its resources.

In a second phase, attention shifted at a global level to Earth's resources as a whole the seas, the air, and biodiversity because of the realization that those essential resources, hitherto taken for granted, were in fact limited. That realization came about through two successive crises. First was the new awareness that humankind had reached a global environmental crisis, which triggered a second wave of bureaucracies to manage nature at both the national and international levels. At the international level the United Nations Environment Programme (UNEP) was established in 1972, and at the national level environmental ministries (in Europe) or agencies (in the United States) flourished in developed countries in the early 1970s. Significantly, the issue that popularized the environmental crisis was an issue of failed global resource management: the overexploitation of whales, which threatened certain species with extinction. That initial awareness was rapidly compounded by a global energy crisis that began in 1973. At that juncture natural resource management was recast as an issue of environmental governance, and it was connected to the new environmental discourses taking shape, such as sustainable development.

The issue of Earth's limits is an unsettled question, not least because of the sheer diversity of natural resources and the difficulty in assessing them scientifically. However, the remaining controversy revolves not around the idea that Earth's resources are limited but rather around its regenerative capacities, which in turn determines how those limits are to be considered. If they are relative, the question becomes one of regulating the access to a resource more stringently or of adapting the activity relying upon it so that it can use a more abundant primary resource. If the limits are absolute, then merely switching the activity from one dependency basis to another is simply insufficient, especially if such dependency continues to expand overall.

Conceptual Approaches to Natural Resource Management

Natural resource management ties in with applied concepts such as maximum sustainable yield (MSY) and optimum utilization. Every natural resource has its optimum utilization, or acceptable levels of use, which are established scientifically and according to which management authorities regulate its exploitation. Such a concept presupposes scientific knowledge as a basis for management and also a regulatory authority (whether national or international) capable of enforcing the exploitation of the resources in accord with such scientific knowledge. The MSY is a regulatory concept that translates precepts of population dynamics into a management tool. Population studies in fisheries have shown that, in a given population, when the deaths increase as a result of human harvesting (exploitation), reproduction rates usually start to rise (as if compensating for the deaths). Theoretically, that resultant surplus production can be harvested sustainably, provided that the harvest is consistently maintained under the MSY, which is specific to each population (rather than to the species as a whole). That is the peak level, beyond which the negative effect of decreasing numbers on the overall population starts to exceed the positive effect of increased reproduction rates. Subsequently, the population as a whole (and not just the harvestable surplus) begins to decline. On the other hand, maintaining exploitation levels below the MSY creates an efficient use of the resources' regenerative capacities, thus in principle enabling exploitation to continue indefinitely. The use of this tool, which was first developed in fisheries, has been extended more broadly, notably through its incorporation into the Convention on the Law of the Sea. However, it has tended to be associated with species-specific management regimes. It requires careful monitoring of population growth and overall health, and overharvesting can easily occur if the population suffers outside declines, such as disease or habitat loss, and harvesting levels are not adjusted. Additional regulations, such as limiting the harvest of females or immature individuals, can help ensure that populations are maintained at sustainable levels.

Another way to think about natural resource management is to think about the purpose of the management. The management objectives are determined by the purpose of the resource itself as a primary resource, as a raw material or fuel, as a source of food, or as a recreational resource. Those uses fall into two broad categories, consumptive and nonconsumptive. Consumptive utilization implies a once-only form of use that is, it refers to activities in which the resource is effectively consumed or used up, such that it cannot be utilized by another party. Hence, the possibility of future exploitation relies on the resource's ability to regenerate itself. Nonconsumptive utilization also uses the resource to generate economic value, but it does so without using up the resource itself. That category encompasses most recreational uses of natural resources. In the case of consumptive uses, management implies balancing exploitation with a respect for the resource's regenerative capacities. In nonconsumptive uses, management centres on regulating the ways in which humans interact with the resource and containing the negative effects of those interactions on the resource. In either case, management is always about resolving a tension between the potentially conflicting objectives of protection and exploitation. Sometimes the use of a resource may change over time or from one part of the globe to the next. The overexploitation of whales is a case in point: whales were initially a primary raw material and fuel in the West, until the mid-20th century, whereas today whales are considered a recreational resource in the West and a food in other areas. This coexistence of different forms of use around the same resource has generated conflicts.

Another strand in natural resource management literature focuses on the difficulties in managing collective resources that is, resources not contained within specific territorial boundaries (such as the sea or air) or resources whose management at the local level has global repercussions, such as forests. According to one line of argument, known as the tragedy of the commons, collective resources lack the incentives inherently built into a privately owned resource to self-limit their exploitation so as to ensure that they will last. In other words, individual users tend to consume as much of the collective resource as possible before others can get to it, resulting in the overall loss of the resource for all. Also at stake in discussion of collective resources is the issue of private versus communal management, a question that goes to the heart of environmental politics.

References

- Types-and-threats-to-natural-resources, environment: eartheclipse.com, Retrieved 17 January, 2020
- What-are-biotic-resources, biology-resources: thebigger.com, Retrieved 13 April, 2020
- Natural-resource-management: britannica.com, Retrieved 05 May, 2020
- Pimentel, David; Pimentel, Marcia (September 2006). "Global environmental resources versus world population growth". Ecological Economics. 59 (2): 195–198. doi:10.1016/j.ecolecon.2005.11.034
- Committee on 21st Century Systems Agriculture (2010). Toward Sustainable Agricultural Systems in the 21st Century. National Academies Press. ISBN 978-0-309-14896-2

5
Techniques for Ecological Conservation

Some of the techniques that are used for ecological conservation are bush regeneration, conservation grazing, phytoremediation, prairie and mangrove restoration, gradsect, controlled burning, etc. This chapter closely examines these techniques of ecological conservation for an extensive understanding of the subject.

Conservation Buffer

Conservation buffers are small areas or strips of land in permanent vegetation, designed to intercept pollutants and manage other environmental concerns. Buffers include: riparian buffers, filter strips, grassed waterways, shelterbelts, windbreaks, living snow fences, contour grass strips, cross-wind trap strips, shallow water areas for wildlife, field borders, alley cropping, herbaceous wind barriers, and vegetative barriers.

Strategically placed buffer strips in the agricultural landscape can effectively mitigate the movement of sediment, nutrients, and pesticides within farm fields and from farm fields. When coupled with appropriate upland treatments, including crop residue management, nutrient management, integrated pest management, winter cover crops, and similar management practices and technologies, buffer strips should allow farmers to achieve a measure of economic and environmental sustainability in their operations. Buffer strips can also enhance wildlife habitat and protect biodiversity.

Benefits of Buffers

Conservation buffers slow water runoff, trap sediment, and enhance infiltration within the buffer. Buffers also trap fertilizers, pesticides, pathogens, and heavy metals, and they help trap snow and cut down on blowing soil in areas with strong winds. In addition, they protect livestock and wildlife from harsh weather and buildings from wind damage. If properly installed and maintained, they have the capacity to:

- Remove up to 50 percent or more of nutrients and pesticides.
- Remove up to 60 percent or more of certain pathogens.
- Remove up to 75 percent or more of sediment.

Conservation buffers reduce noise and odor. They are a source of food, nesting cover, and shelter for many wildlife species. Buffers also provide connecting corridors that enable wildlife to move safely from one habitat area to another.

Conservation buffers help stabilize a stream and reduce its water temperature. Buffers also offer a setback distance for agricultural chemical use from water sources.

Like the trim on a house makes the house look better, well-planned conservation buffers improve the appearance of a farm or ranch. If used as part of a comprehensive conservation system, buffers will make good use of areas that often should not be cropped.

Importance of Conservation Buffers

- Conservation buffers work environmentally because they:
    - Control soil erosion by both wind and water.
    - Improve soil quality.
    - Improve water quality by removing sediment, fertilizers, pesticides, pathogens, and other potential contaminants from runoff.
    - Enhance fish and wildlife habitat.
    - Reduce flooding.
    - Conserve energy.
    - Protect buildings, roads, and livestock.
    - Conserve biodiversity.

- Conservation buffers work economically because of financial incentives available through USDA conservation programs the continuous Conservation Reserve Program (CRP) sign-up, Environmental Quality Incentives Program (EQIP), Wildlife Habitat Incentives Program (WHIP), general CRP, Wetlands Reserve Program (WRP), and Conservation Stewardship Program (CSP).
- Conservation buffers may help you meet Federal, state, or local pollution control requirements.
- Many state and local governments and even some private organizations offer additional financial incentives to install conservation buffers.
- Conservation buffers are a visual demonstration of your commitment to land stewardship.

Bush Regeneration

Bush regeneration, a form of natural area restoration, is the term used in Australia for the ecological restoration of remnant vegetation areas, such as through the minimisation of negative disturbances, both exogenous such as exotic weeds and endogenous such as erosion. It may also attempt to recreate conditions of pre-European arrival, for example by simulating endogenous disturbances such as fire. Bush regeneration attempts to protect and enhance the floral biodiversity in an area by providing conditions conducive to the recruitment and survival of native plants. Bushcare's Major Day Out is an Australian national day of community participation in the care of bushland. In 2012 nearly 100 bushcare sites participated in this annual event.

Purposes

The aim of bush regeneration, also known as 'natural area restoration', is to restore and maintain ecosystem health by facilitating the natural regeneration of indigenous flora, this is usually achieved by selectively reducing the competitive interaction with invasive species, or mitigation of negative influences such as weeds or erosion.

Invasive plant species are often the greatest threat to remnant vegetation, and therefore bush regeneration is closely associated with weed abatement activities. Weed management as one aim of bush regeneration, is used to increase native plant recruitment. The management of factors such as fire and herbivory can be just as important, depending on the ecosystem being restored. In recent years research and on-ground management has begun to recognize the importance of ecosystem processes rather ecosystem composition and structure and research into other ways of facilitating native plant recruitment is increasing.

Technique

The original Bradley method of bush regeneration focuses on facilitating native plant recruitment from the seedbank, rather than planting seedlings or sowing seeds, as follows:

"Weeding a little at a time from the bush towards the weeds takes the pressure off the natives under favourable conditions. Native seeds and spores are ready in the ground and the natural environment favours plants that have evolved in it. The balance is tipped back towards regeneration. Keep it that way, by always working where the strongest area of bush meets the weakest weeds".

Currently the term 'bush regeneration' includes activities other than weed removal, such as replanting and introducing species into an area where soil, water, or fire regimes have shifted the type of plant appropriate to the area (e.g. a stormwater drain).

Weed species can be important habitat for native fauna (e.g. blackberry is important habitat for wrens and the southern brown bandicoot) and this should be taken into consideration with bush regeneration, for example by not clearing invasive species until adequate habitat alternatives have been established nearby with native vegetation.

Problems can occur when insufficient follow-up is conducted as the success of bush regeneration is dependent on allowing the native vegetation to regenerate in the area where weeds have been removed.

Dew Pond

Typical example of downland dew pond near Chanctonbury Ring, West Sussex.

A dew pond is an artificial pond usually sited on the top of a hill, intended for watering livestock. Dew ponds are used in areas where a natural supply of surface water may not be readily available. Despite the name, their primary source of water is believed to be rainfall rather than dew or mist.

Construction

Modern examples made with portland cement need regular repair. Oxteddle Bottom, Sussex, England.

They are usually shallow, saucer-shaped and lined with puddled clay, chalk or marl on an insulating straw layer over a bottom layer of chalk or lime. To deter earthworms from their natural tendency of burrowing upwards, which in a short while would make the clay lining porous, a layer of soot would be incorporated or lime mixed with the clay. The clay is usually covered with straw to prevent cracking by the sun and a final layer of chalk rubble or broken stone to protect the lining from the hoofs of sheep or cattle. To retain more of the rainfall, the clay layer could be extended across the catchment area of the pond.

If the pond's temperature is kept low, evaporation (a major water loss) may be significantly reduced, thus maintaining the collected rainwater.This may be attained by building the pond in a hollow, where cool air is likely to gather, or by keeping the surrounding grass long to enhance heat radiation. As the water level in the basin falls, a well of cool, moist air tends to form over the surface, restricting evaporation.

Measuring of Dew Production

At the dew pond of Helmfleeth in the municipality Poppenbüll (Peninsula Eiderstedt in Schleswig-Holtstein, Germany), in situ measurements of the evaporation and condensation were taken. For this reason, meteorological measuring instruments and a floating evaporation pan were used. These measurements proved the dew formation on the basis of temperature changes and the weather conditions. The dew pond Helmfleeth is part of the water supply of marsh areas and is used until today.

Instruments measuring condensation and evaporation at a Helmfleeth dew pond.

Reproductions of Historical Dew Ponds

In 2014, the traditional technique was verified by means of modern building material at reproductions of dew ponds in East Friesland. In this context, various techniques were tried in two terrestrial hollows. Commercially available PVC-film was used for the sealing and foam glass gravel for the insulation. The construction was carried out by craftsmen and the climatological analysis by Werner and Coldewey.

Gradsect

A gradsect or gradient-directed transect is a low-input, high-return sampling method where the aim is to maximise information about the distribution of biota in any area of study. Most living things are rarely distributed at random, their placement being largely determined by a hierarchy of environmental factors. For this reason, standard statistical designs based on purely random sampling or systematic (e.g. grid-based) systems tend to be less efficient in recovering information about the distribution of taxa than sample designs that are purposively directed instead along deterministic environmental gradients.

Ecologists have long been aware of the significance of environmental gradient based approaches to better understand community dynamics. Although in practice, life-scientists intuitively sample gradients, until the early 1980s there was little formal theoretical or empirical support for such an approach, sample design being driven largely by traditional statistical methods based on probability theory incorporating random sampling.

Intensively sampled landscape-based surveys in Australia provided a reference platform for developing and testing a less logistically demanding and yet statistically acceptable gradient-based survey design that avoided the need for random or purely grid-based sampling. These initial studies and subsequently developed statistical support for purposive, gradient-based survey provided a formalized, practical alternative to more logistically demanding traditional designs. It was here the term gradsect was coined that coupled purposive, transect sampling with a hierarchical framework of environmental gradients considered to be key determinants of species distribution.

Methodology

In constructing a gradsect, existing information is initially reviewed in which a hierarchy of environmental gradients is first identified either by visual means (maps, aerial photographs etc..) or through numerical analysis or spatial analysis of institutional or other data sources. A typical regional gradsect for example may be constructed according to a primary climate gradient (temperature, moisture, seasonality) then a secondary gradient (geomorphology, lithology, major and minor drainage systems), a tertiary gradient possibly represented by a local soil catena or local land use farming system or finer scale gradient levels representing local vegetational sequences. Through an inspection of spatial overlays of all gradients, a minimum number of sample locations is then purposively located to reflect, as far as possible, total environmental variation. For logistic and other purposes (such as improving the capacity to locate rare species) the steepest gradients are usually selected. In this way an ideal gradsect is constructed that may then be modified to accommodate logistic tradeoffs. The selection discipline requires that the fullest possible range of each hierarchical level is sampled. This commonly results in a set of progressively nested clusters of sample sites contained within the overarching primary gradient that may not reflect a linear distribution. At relatively local landscape scale, a primary gradients may be represented by salinity levels or water depth as in tidal wetlands or micro-topographic relief as in forest margins or a riparian zone. For most practical purposes, transects are commonly laid out along contours perpendicular to the main direction of the gradient. Iterative spatial analysis of environmental layers over a digital elevation model can then be used to identify areas requiring additional sampling thereby improving environmental representativeness.

Advantages and Limitations

Initial studies in gradsect development revealed considerable logistic and other advantages over more traditional non-gradient-based survey designs concerned primarily with random sampling. This finding is now widely supported especially in biodiversity and other areas of environmental surveying and conservation design. Apart from improved logistic efficiency, the gradsect method seeks to maximise environmental representativeness which has the dual advantage of potentially improving location of rarities and enhancing spatial modelling of species distribution. Because the underlying statistical model is not based on probability theory, gradsect sampling cannot be used to estimate numbers of species or other biological attributes per unit area. For that purpose some measure of random sampling needs to be built into the sample design.

Applications

Since the publication of gradsect theory in 1984, subsequent vegetational and landscape studies in regional Australia; Ludwig and Tongway were followed by a successful evaluation of the method in faunal surveys in South Africa. Since then applications involving gradsects have ranged from habitat suitability studies of fungi, termites other macro invertebrates ; birds small and large mammals. Vegetation studies using gradsects have been widely applied in many countries ranging from tidal wetlandsand agricultural cropping systems and forested landscape mosaics to infectious diseases. At broader geographic and national scales gradsects have been applied to guide field sampling and forest mapping in mountainous terrain as well as wide-ranging remote sensing applications.

Mark and Recapture

Collar tagged rock hyrax.

Jackdaw with a numbered aluminum ring on its left tarsus.

Mark and recapture is a method commonly used in ecology to estimate an animal population's size. A portion of the population is captured, marked, and released. Later, another portion will be captured and the number of marked individuals within the sample is counted. Since the number of marked individuals within the second sample should be proportional to the number of marked individuals in the whole population, an estimate of the total population size can be obtained by dividing the number of marked individuals by the proportion of marked individuals in the second sample. The method is most useful when it is not practical to count all the individuals in the population. Other names for this method, or closely related methods, include capture-recapture, capture-mark-recapture, mark-recapture, sight-resight, mark-release-recapture, multiple systems estimation, band recovery, the Petersen method, and the Lincoln method.

Biologist is marking a Chittenango ovate amber snail to monitor the population.

Marked Chittenango ovate amber snail.

Another major application for these methods is in epidemiology, where they are used to estimate the completeness of ascertainment of disease registers. Typical applications include estimating the number of people needing particular services (i.e. services for children with learning disabilities, services for medically frail elderly living in the community), or with particular conditions (i.e. illegal drug addicts, people infected with HIV, etc.).

Typically a researcher visits a study area and uses traps to capture a group of individuals alive. Each of these individuals is marked with a unique identifier (e.g., a numbered tag or band), and then is released unharmed back into the environment. A mark recapture method was first used for ecological study in 1896 by C.G. Johannes Petersen to estimate plaice, Pleuronectes platessa, populations.

Sufficient time is allowed to pass for the marked individuals to redistribute themselves among the unmarked population.

Next, the researcher returns and captures another sample of individuals. Some individuals in this second sample will have been marked during the initial visit and are now known as recaptures. Other animals captured during the second visit, will not have been captured during the first visit to the study area. These unmarked animals are usually given a tag or band during the second visit and then are released.

Population size can be estimated from as few as two visits to the study area. Commonly, more than two visits are made, particularly if estimates of survival or movement are desired. Regardless of the total number of visits, the researcher simply records the date of each capture of each individual. The "capture histories" generated are analyzed mathematically to estimate population size, survival, or movement.

Notation

Let,

N = Number of animals in the population.

n = Number of animals marked on the first visit.

K = Number of animals captured on the second visit.

k = Number of recaptured animals that were marked.

A biologist wants to estimate the size of a population of turtles in a lake. She captures 10 turtles on her first visit to the lake, and marks their backs with paint. A week later she returns to the lake and captures 15 turtles. Five of these 15 turtles have paint on their backs, indicating that they are recaptured animals. This example is $(n, K, k) = (10, 15, 5)$. The problem is to estimate N. N=n*K/k.

Lincoln–Petersen Estimator

The Lincoln–Petersen method (also known as the Petersen–Lincoln index or Lincoln index) can be used to estimate population size if only two visits are made to the study area. This method assumes that the study population is "closed". In other words, the two visits to the study area are close enough in time so that no individuals die, are born, or move into or out of the study area between visits. The model also assumes that no marks fall off animals between visits to the field site by the researcher, and that the researcher correctly records all marks.

Given those conditions, estimated population size is:

$$\hat{N} = \frac{Kn}{k}.$$

Derivation

It is assumed that all individuals have the same probability of being captured in the second sample, regardless of whether they were previously captured in the first sample (with only two samples, this assumption cannot be tested directly).

This implies that, in the second sample, the proportion of marked individuals that are caught (k / K) should equal the proportion of the total population that is marked (n / N). For example, if half of the marked individuals were recaptured, it would be assumed that half of the total population was included in the second sample.

$$\frac{k}{K} = \frac{n}{N}.$$

A rearrangement of this gives.

$$\hat{N} = \frac{Kn}{k},$$

the formula used for the Lincoln–Petersen method.

Sample Calculation

In the example (n, K, k) = (10, 15, 5) the Lincoln–Petersen method estimates that there are 30 turtles in the lake.

$$\hat{N} = \frac{Kn}{k} = \frac{10 \times 15}{5} = 30$$

Chapman Estimator

The Lincoln–Peterson estimator is asymptotically unbiased as sample size approaches infinity, but is biased at small sample sizes. An alternative less biased estimator of population size is given by the Chapman estimator:

$$\hat{N}_C = \frac{(K+1)(n+1)}{k+1} - 1$$

Sample Calculation

The example (K, n, k) = (10, 15, 5) gives:

$$\hat{N}_C = \frac{(K+1)(n+1)}{k+1} - 1 = \frac{11 \times 16}{6} - 1 = 28.3$$

Note that the answer provided by this equation must be truncated not rounded. Thus, the Chapman method estimates 28 turtles in the lake.

Surprisingly, Chapman's estimate was one conjecture from a range of possible estimators: "In practice, the whole number immediately less than (K+1)(n+1)/(k+1) or even Kn/(k+1) will be the estimate. The above form is more convenient for mathematical purposes. Chapman also found the estimator could have considerable negative bias for small Kn/N, but was unconcerned because the estimated standard deviations were large for these cases.

Confidence Interval

An approximate $100(1-\alpha)\%$ confidence interval for the population size N can be obtained as:

$$K+n-k-0.5+\frac{(K-k+0.5)(n-k+0.5)}{(k+0.5)}\exp(\pm z_{\alpha/2}\hat{\sigma}_{0.5}),$$

where $z_{\alpha/2}$ corresponds to the $1-\alpha/2$ quantile of a standard normal random variable, and

$$\hat{\sigma}_{0.5}=\sqrt{\frac{1}{k+0.5}+\frac{1}{K-k+0.5}+\frac{1}{n-k+0.5}+\frac{k+0.5}{(n-k+0.5)(K-k+0.5)}}$$

It has been shown that this confidence interval has actual coverage probabilities that are close to the nominal $100(1-\alpha)\%$.level even for small populations and extreme capture probabilities (near to 0 or 1), in which cases other confidence intervals fail to achieve the nominal coverage levels.

Bayesian Estimate

The mean value ± standard deviation is,

$$N \approx \mu \pm \sqrt{\mu \grave{o}}$$

where,

$$\mu=\frac{(K-1)(n-1)}{k-2} \text{ for } k>2$$

$$\epsilon=\frac{(K-k+1)(n-k+1)}{(k-2)(k-3)} \text{ for } k>3$$

The example (*K*, *n*, *k*) = (10, 15, 5) gives the estimate $N \approx 42 \pm 21.5$.

Capture Probability

Bank vole, Myodes glareolus, in a capture-release small mammal population study for London Wildlife Trust at Gunnersbury Triangle local nature reserve.

The capture probability refers to the probability of a detecting an individual animal or person of interest, and has been used in both ecology and epidemiology for detecting animal or human diseases, respectively.

The capture probability is often defined as a two-variable model, in which f is defined as the fraction of a finite resource devoted to detecting the animal or person of interest from a high risk sector of an animal or human population, and q is the frequency of time that the problem (e.g., an animal disease) occurs in the high-risk versus the low-risk sector. For example, an application of the model in the 1920s was to detect typhoid carriers in London, who were either arriving from zones with high rates of tuberculosis (probability q that a passenger with the disease came from such an area, where q>0.5), or low rates (probability 1-q). It was posited that only 5 out of every 100 of the travelers could be detected, and 10 out of every 100 were from the high risk area. Then the capture probability P was defined as:

$$P = \frac{5}{10} fq + \frac{5}{90}(1-f)(1-q),$$

where the first term refers to the probability of detection (capture probability) in a high risk zone, and the latter term refers to the probability of detection in a low risk zone. Importantly, the formula can be re-written as a linear equation in terms of f:

$$P = (\frac{5}{10}q - \frac{5}{90}(1-q))f + \frac{5}{90}(1-q).$$

Because this is a linear function, it follows that for certain versions of q for which the slope of this line (the first term multiplied by f) is positive, all of the detection resource should be devoted to the high-risk population (f should be set to 1 to maximize the capture probability), whereas for other value of q, for which the slope of the line is negative, all of the detection should be devoted to the low-risk population (f should be set to 0. We can solve the above equation for the values of q for which the slope will be positive to determine the values for which f should be set to 1 to maximize the capture probability:

$$\left(\frac{5}{10}q - \frac{5}{90}(1-q)\right) > 0,$$

which simplifies to:

$$q > \frac{1}{10}.$$

This is an example of linear optimization. In more complex cases, where more than one resource f is devoted to more than two areas, multivariate optimization is often used, through the simplex algorithm or its derivatives.

Integrated Approaches

Modelling mark-recapture data is trending towards a more integrative approach, which combines mark-recapture data with population dynamics models and other types of data. The integrated approach is more computationally demanding, but extracts more information from the data improving parameter and uncertainty estimates.

Controlled Burn

Controlled burning is any fire intentionally ignited to meet specific land management objectives, such as to reduce flammable fuels, restore ecosystem health, recycle nutrients, or prepare an area for new trees or vegetation. Controlled burning is a management tool that when used under specifically controlled conditions will help land stewards manage forests and rangelands for multiple use.

Land management agencies have done a tremendous job educating the public about the dangers of wildland fire and reducing the number of human caused fires. However, it is important to realize that not all fire is bad. In fact, many of our ecosystems are dependent on fire.

- Fire historically crept through these areas with low intensity.
- Every 5 to 15 years, fire regenerated the forest and cleansed the understory of potential hazardous fuels. These historic fires were not the devastating wildfires of recent years.

- Frequent cool fires acted as a natural agent reducing surface fuels and all but eliminated large, stand replacing, fire events that have become too frequent during the last three decades.

Fire exclusion practices have resulted in forests being plagued with a variety of problems, including overcrowding resulting from encroachment of species normally eliminated by fire; vulnerability of trees to insects and disease; and inadequate reproduction of fire resistant species. In addition, heavy accumulation of fuel dead vegetation of forest floors can cause catastrophic fires, threaten public safety, impair forests and ecosystem health, and degrade air quality.

The Forest Service has made it a priority to reintroduce fire into fire dependent ecosystems to help promote ecosystem health. Controlled burning is viewed by the land management agencies as an agent of change that helps "mother nature" return an ecosystem to its historic range.

Understanding fire is a science. The ability to know when an ecosystem is ready for controlled burning is science. When a land management agency restores fire to a given area, they first must define the objectives that the fire would achieve. There are also many elements of nature that must be just right to meet the objectives of controlled burning, from both dead and live fuel moistures, air temperature, to wind speed and humidity. This is referred to as the "burn window," the preplanned condition targeted for burning. Due to the uncertainty of these natural elements, fire specialists monitor the critical elements of nature.

Controlled Burning and Air Quality

How fire affects air quality depends on many factors, including weather conditions, such as wind and humidity; the scope and severity of the fire; and the type and quantity of fuels burned.

Smoke contains a number of pollutants. Tiny particles created from the smoke is a concern because it can cause serious health problems. Smoke also adversely affects the clarity of our air, which in turn, affects the distance and sharpness of what we see.

Unplanned or unwanted fires can pose serious threats to public health and safety, as well as to air quality. Because these fires are uncontrolled, they can pose significant threats to the safety of firefighters and the general public, as well as destroy property.

On the other hand, planned or managed fires, are used to minimize the emissions and adverse impacts of smoke on public health and the environment. Many techniques are used to manage the impacts from smoke, including scheduling burning during favorable weather conditions and controlling the amount of fuel and acreage burned. In addition, all planned fire activities are already subject to state air quality regulations.

Fuels in your National Forest

Fire has been a natural part of this forest's ecosystem for a very long time. Throughout time wildfires ignited and burned naturally through the forest. Some were caused by lightning and some were intentionally started by Native Americans. These low intensity fires in the past kept the forest floor free from the natural annual build up of tree needles, dead grass, thick brush, and dead trees. As a result, fire has shaped vegetation patterns and wildlife distributions in the national forests.

The next time you are camping, hiking or driving in your national norest notice how much limb wood, brush, and dense thickets of small trees cover the forest floor. This build up of fuel feeds wildfires. Today, fires that are not caught when they are small quickly build in size and fire intensity into catastrophic wildfires.

The Forest Service uses controlled burning as its way to put fire back into the ecosystem.

Nature's Way of Recycling

Fallen trees and limbs left to rot on the forest floor will decay at a very slow rate. In fact, large logs can take more than 100 years to decompose. This process is aided by the numerous species of bacteria, insects, and wildlife that live in the decomposing materials. All this rotting is one way that nature recycles nutrients back into the soil. Pine needles decompose very slowly. It takes more than a year for 10% of the pine needles to decay. As a result, year after year, pine needles continue to build up until they are eliminated by fire.

Faster recycling occurs during a fire. Gasses are released into the atmosphere in the form of smoke. In the burned area, nitrogen and other nutrients remain and are leached back into the soil as rain soaks the ground. This is nature's way of rapidly recycling nutrients. Unfortunately, when there's too much fuel on the ground and it's burned in an intense wildfire these benefits are often missing. Intense fires tend to scorch the ground and kill the trees above.

Which is Better, Slow or Fast Recycling?

Slowly decomposing materials release nutrients steadily into the soil. This continuous release helps to maintain growth over a long period of time. Decomposition can last indefinitely as long as dead material on the forest floor continues to accumulate. In other areas of the world, a wildfire may happen once every 600 years. In these areas, only decomposition will supply needed nutrients between wildfires.

Wildfires cause fast nutrient recycling. An abundant supply of nutrients helps new seedlings, brush, and grasses to grow quickly and become established following a wildfire. This is nature's way of quick starting a forest. Most of the nutrients are quickly used up; however, more lasting effects occur when they are leached into the soil. It usually takes a few years for the supply of nutrients to return to normal levels.

The forest needs both slow recycling, from decomposition, and fast recycling from fires. However, fires ignited at the wrong time can quickly turn into catastrophic events.

Wildfires occurred every seven to 25 years in Kentucky's forests prior to the early 1900s. Around 1915 national forest management policy viewed wildfires as destructive and began putting them all out as quickly as possible. Today we realize the important role fire plays in maintaining healthy national forests.

Why not use Controlled Burning Everywhere?

Some places are not easy to conduct a controlled burn. These are locations where it may not be economical, feasible, or practical. Exact fire prescriptions are developed by fire managers before burning is allowed. These fire prescriptions are based on weather, moisture content of the fuels, and how the fire can be lighted (ignition patterns). There may only be 50 days in an entire year when an area meets the prescription.

Laws and regulations also determine when a controlled burn may be ignited. Air quality regulations play an important role. It's not unusual for forest conditions to be in prescription and a no burn day due to poor air quality. Controlled burning takes place when laws, regulations, and forest needs are all in balance. Controlled burns also don't take place when there are safety, health, and aesthetic concerns.

The ecosystem of this forest was developed with fire as one of the main sculptors of vegetation found there. Fire related processes need to continue to perpetuate this ecosystem.

Ecological Restoration

Ecological restoration aims to recreate, initiate, or accelerate the recovery of an ecosystem that has been disturbed. Disturbances are environmental changes that alter ecosystem structure and function. Common disturbances include logging, damming rivers, intense grazing, hurricanes, floods, and fires. Restoration activities may be designed to replicate a pre-disturbance ecosystem or to create a new ecosystem where it had not previously occurred. Restoration ecology is the scientific study of repairing disturbed ecosystems through human intervention.

Goals

Restoration projects differ in their objectives and their methods of achieving those goals. Many restoration projects aim to establish ecosystems composed of a native species; other projects attempt to restore, improve, or create particular ecosystem functions, such as pollination or erosion control. Some examples of different kinds of restoration include the following:

- Revegetation: The establishment of vegetation on sites where it has been previously lost, often with erosion control as the primary goal. For example, vegetated buffers are strips of vegetation that protect water quality in riparian ecosystems from urban or agricultural runoff.

- Habitat enhancement: The process of increasing the suitability of a site as habitat for some desired species.
- Remediation: Improving an existing ecosystem or creating a new one with the aim of replacing another that has deteriorated or been destroyed.
- Mitigation: Legally mandated remediation for loss of protected species or ecosystems.

At a given restoration site, it may be possible to establish a number of different communities. When choosing a target state for a restoration project however, restorationists generally select only one (or a small range) of possible community types. Often some sort of "natural" pre-disturbance condition, or reference state, is selected, along with its presumed properties (e.g., previous flooding or fire patterns). This is often represented by a nearby undisturbed reference site. Even with good working knowledge of an historical ecosystem's species composition and functions, practitioners must still decide how far in the past is defined as "natural." For some ecosystems the reference state may be before any human disturbance, but more commonly the reference state is before agricultural or industrial intensification (such as pre-European contact in the US). However, sometimes an historical target may no longer be appropriate under current or projected climatic or biotic conditions. For example, future climates may not support certain species, and some species may have already gone extinct in an area. Under these circumstances practitioners may decide to create an ecosystem that did not exist historically at the project site, but which corresponds to current or projected future conditions. Sometimes, restoration efforts are designed to maintain a desirable human-derived state, such as montane meadows or Scottish moors.

Concepts Underpinning Restoration

Disturbance

Disturbance events can occur at many scales and different levels of severity, and some are natural parts of every ecosystem. Disturbance events can alter species composition, nutrient cycling, and soil properties. Natural disturbances include severe weather damage, fire, flooding, treefalls and even volcanic eruptions. Anthropogenic (human-caused) disturbances can alter or destroy natural habitat (like clearing land for agriculture) and/or ecological functions (like damming rivers for flood control). Humans can also change natural disturbance events and cycles (like suppression of wildfires and prevention of periodic flooding). The goal of a restoration project may be to initiate or speed the recovery of an ecosystem after disturbance. Restoration activities may also be designed to reestablish natural disturbance regimes.

Genetics

Restoration projects also typically include genetic considerations. Plants (or animals) from local sources are more likely to be well adapted to the target ecosystem. Therefore, using animals or plant materials (like seeds or cuttings) collected from local sources may increase the chance of successful establishment. Including a large number of individual plants or animals can help ensure genetic diversity in the restored populations. Genetic diversity is thought to be critical to maintaining the ability of populations to evolve and recover from disturbances.

Succession

Ecological succession is the process by which biological community composition- the number and proportion of different species in an ecosystem- recover over time following a disturbance event. Passive restoration means simply allowing natural succession to occur in an ecosystem after removing a source of disturbance. The recovery of the deciduous forests in the eastern United States after the abandonment of agriculture is a classic example of passive restoration. Active restoration involves accelerating the process or attempting to change the trajectory of succession. For example, mine tailings would take so long to recover passively that active restoration is usually appropriate.

Community Assembly Theory

Community assembly theory suggests that similar sites can develop different biological communities depending on order of arrival of different species. In the context of restoration, sites may not always recover toward a desired or anticipated group of species or ecosystem functions. Composition of seed mixes, planting order and year of planting may be important considerations for restoration practitioners, particularly when goals include the establishment of certain ecological communities or the prevention of invasion by weeds or pests.

Landscape Ecology

Restoration draws on several concepts from landscape ecology. Restored areas are often relatively small and isolated, which makes them especially sensitive to problems associated with habitat fragmentation. Habitat fragmentation occurs when continuous areas of habitat become disconnected by natural or human causes (for example, building roads through a forest). Fragmentation generally leads to small, isolated patches of hospitable habitat. Smaller habitats support fewer species and smaller populations, which are at greater risk of inbreeding and local extinction. The theory of island biogeography predicts that populations are more likely to persist in habitat patches that are large and/or well connected with populations in other hospitable habitats. This theory assumes that the matrix the region between habitat patches is uniform and inhospitable. The most common examples of this concept are oceanic islands, dots of terrestrial species' habitat surrounded by uninhabitable water. More recently, the classic dichotomy of hospitable versus inhospitable habitat has been modified to include the existence a multiple types of habitat patches which are juxtaposed to form a patch mosaic. These different patches within the mosaic may be more or less hospitable for the species, communities and ecosystem functions targeted by restoration activities.

Fragmentation may also intensify negative edge effects impacts of one habitat on an adjacent habitat by increasing the amount of edge habitat and reducing the distances among edges. For instance, invasive weeds are more abundant along forest edges, so small forest fragments (which have more edge habitat) are more likely to be invaded. Restoration activities often seek to improve connectivity among habitat patches in fragmented landscapes by creating or restoring linkages. Examples of linkages commonly used to improve connectivity are corridors and stepping stones. Corridors are relatively narrow, linear strips of habitat between otherwise isolated habitat patches. Stepping stones are small unconnected patches of habitat that are close enough together to allow movement across the landscape.

Application

Applied restoration is a multi-step process, which may include some or all of these stages:

- Assessing the site: A thorough appraisal of the current conditions at the restoration site is essential for determining what kind of actions will be necessary. In this step, the causes of ecosystem disturbance and methods for stopping or reversing them are identified.
- Formulating project goals: To determine targets for the restored community, practitioners may visit reference sites (similar, nearby environments in natural condition) and/or consult historical sources that detail the pre-disturbance community. Goals may also include considerations of what species will be best suited to present or future climate conditions.
- Removing sources of disturbance: Before restoration can be successful, forces of disturbance may need to be removed. Examples include cessation of mining or farming or causes of erosion, restricting livestock from riparian areas, removing toxic materials from soil or sediments, and eradicating invasive exotic species.
- Restoring processes/disturbance cycles: Sometimes restoring important ecological processes such as natural flood or fire regimes is enough to restore ecosystem integrity. In these cases, native plants and animals that have evolved to tolerate or require natural disturbance regimes may come back on their own without direct action by practitioners.

- Rehabilitating substrates: This can include any activity aimed at repairing altered soil texture or chemistry, or restoring hydrological regimes or water quality.
- Restoring vegetation: In many cases, restoration activities involve direct revegetation of a site. Usually, native species suited to local environmental conditions are chosen for planting. Seeds or cuttings are generally collected from a variety of sources within a local region in order to ensure genetic diversity. Vegetation can be planted as seeds, or seedlings.
- Monitoring and maintenance: Monitoring the restoration site over time is critical to determining whether goals are being met, and can inform future management decisions. Observations made at the site may indicate that further action, such as periodic weed removal, is necessary in ensuring the long-term success of the project. Ideally restoration projects would eventually achieve a self-sustaining ecosystem without the need for future human intervention.

Virtually all the worlds' ecosystem types have been the subject of restoration efforts, but particular attention has been paid to ecosystems most impacted by human activities, such as wetlands, grasslands/rangelands, riparian areas, and tropical forests.

Agroecological Restoration

Agroecological restoration is the practice of re-integrating natural systems into agriculture in order to maximize sustainability, ecosystem services, and biodiversity. This is one example of a way to apply the principles of agroecology to an agricultural system.

Farms cannot be restored to a purely natural state because of the negative economic impact on farmers, but returning processes, such as pest control to nature with the method of intercropping, allows a farm to be more ecologically sustainable and, at the same time, economically viable. Agroecological restoration works toward this balance of sustainability and economic viability because conventional farming is not sustainable over the long run without the integration of natural systems and because the use of land for agriculture has been a driving force in creating the present world biodiversity crisis. Its efforts are complementary to, rather than a substitute for, biological conservation.

Agriculture creates a conflict over the use of land between wildlife and humans. Though the domestication of crop plants occurred 10,000 years ago, a 500% increase in the amount of pasture and crop land over the last three hundred years has led to the rapid loss of natural habitats. In recent years, the world community acknowledged the value of biodiversity in treaties, such as the 1992 landmark Convention on Biological Diversity.

Reintegration

The reintegration of agricultural systems into more natural systems will result in decreased yield and produce a more complex system, but there will be considerable gains in biodiversity and ecosystem services.

Biodiversity

The Food and Agriculture Organization of the United Nations estimates that more than 40% of earth's land surface is currently used for agriculture. And because so much land has been converted to agriculture, habitat loss is recognized as the driving force in biodiversity loss (FAO). This biodiversity loss often occurred in two steps, as in the American Midwest, with the introduction of mixed farming carried out on small farms and then with the widespread use of mechanized farming and monoculture beginning after World War II. The decline in farmland biodiversity can now be traced to changes in farming practices and increased agricultural intensity.

Increasing Heterogeneity

Heterogeneity (here, the diversity or complexity of the landscape) has been shown to be associated with species diversity. For example, the abundance of butterflies has been found to increase with heterogeneity. One important part of maintaining heterogeneity in the spaces between different fields is made up of habitat that is not cropped, such as grass margins and strips, scrub along field boundaries, woodland, ponds, and fallow land. These seemingly unimportant pieces of land are crucial for the biodiversity of a farm. The presence of field margins benefits many different taxa: the plants attract herbivorous insects, will which attract certain species of birds and those birds will attract their natural predators. Also, the cover provided by the no cropped habitat allows the species that need a large range to move across the landscape.

Monoculture

In the absence of cover, species face a landscape in which their habitat is greatly fragmented. The isolation of a species to a small habitat that it can't safely wander from can create a genetic bottleneck, decreasing the resilience of the particular population, and be another factor leading to the decline of the total population of the species. Monoculture, the practice of producing a single crop over a wide area, causes fragmentation. In conventional farming, monoculture, such as with rotations of corn and soybean crops planted in alternating growing seasons, is used so that very high yields can be produced. After the mechanization of farming, monoculture became a standard practice in corn-beans rotation, and had broad implications for the long-term sustainability and biodiversity of farms. Whereas organic fertilizers, had kept the soil's nutrients fixed to the ecosystem, the introduction of monoculture removed the nutrients and farmers compensated for that loss by using inorganic fertilizers. It is estimated that humans have doubled the rate of nitrogen input into the nitrogen cycle, mostly since 1975. As a result, the biological processes that controlled the way crops used the nutrients changed and the leached nitrogen from farmland soils has become a source of pollution.

Organic Farming

Organic farming prohibits the use of synthetic chemicals in crop and livestock production. It often includes diverse crop rotations and provides non-cropped habitat for insects that provide ecosystem services, such as pest control and pollination. Organic farmers are encouraged to follow wildlife friendly practices, and as a result there is a great difference between the ecosystem services that similarly sized but distinctly managed organic farms provide. A review of the 76 studies concerning the relationship between biodiversity and organic farming listed three practices associated with organic farming that accounted for the higher biodiversity counts found in organic farms as compared to conventional farms.

- Prohibition/reduced use of chemical pesticides and inorganic fertilizers is likely to have a positive impact through the removal of both direct and indirect negative effects on arable plants, invertebrates and vertebrates.
- Sympathetic management of non-crop habitats and field margins can enhance diversity and abundance of arable plants, invertebrates, birds and mammals.
- Preservation of mixed farming is likely to positively impact farmland biodiversity through the provision of greater habitat heterogeneity at a variety of temporal and spatial scales within the landscape.

Conservation Grazing

Highland Cattle on the Grazing Marsh at London Wetland Centre.

Conservation grazing or targeted grazing is the use of semi-feral or domesticated grazing livestock to maintain and increase the biodiversity of natural or semi-natural grasslands, heathlands, wood pasture, wetlands and many other habitats. Conservation grazing is generally less intensive than practices such as prescribed burning, but still needs to be managed to ensure that overgrazing does not occur. The practice has proven to be beneficial in moderation in restoring and maintaining grassland and heathland ecosystems. The optimal level of grazing will depend on the goal of conservation, and different levels of grazing, alongside other conservation practices, can be used to induce the desired results.

Conservation Grazing in Practice

Urban ecopastoralism on a Historical Monument, the Citadel of Lille, by Soay sheep. Their long coats help to disperse plant seeds.

Urban ecopastoralism with sheep and goats in an urban meadow of the "Bois de la Citadelle" in Lille.

Grazing maintains an area as a habitat dominated by grasses and small shrubs, largely preventing ecological succession to forest.

Conservation grazing needs to be monitored closely. Overgrazing may cause erosion, habitat destruction, soil compaction, or reduced biodiversity (species richness). *Rambo and Faeth* found that the use of vertebrates for grazing of an area increased the species richness of plants by decreasing the abundance of dominant species and increasing the richness of rarer species. This may lead to a more open forest canopy and more room for other plant species to emerge.

Conservation grazing Longhorn Cattle to manage national nature reserve at Ruislip Lido.

Grazing Restoration Effect Dependent on Grazer Species

Different grazing species have different effects. For example, elk and horses have a similar grazing frequency to cattle but tend to spread their zone of grazing to cover a greater area, producing a smaller effect on a given area than cattle would. Similarly, cattle have been found to be more useful in the restoration of pastures with low species richness, and sheep were found useful for the re-establishment of neglected fields. The type of area that needs to be restored or maintained will determine the species of grazer ideal for conservation grazing. In the use of varied breeds of steers, "traditional breeds appeared slightly less selective than commercial breeds", but did not make a significant difference in biodiversity. In this particular study biodiversity was maintained by the same amount by both breed types.

Effects on Native and Non-native Plant Species

Conservation grazing is a tool used for conserving biodiversity. However, one danger in grazing is the potential for invasive species to be enhanced as well as the native biodiversity. Areas of high intensity grazing and grazer removal increased the biomass of nonnative introduced species. Both showed that an intermediate approach is the best method. The nonnatives did demonstrate that they were not as well adapted to the disturbances, such as drought. This indicated that implementing controlled grazing methods would decrease the abundance of nonnatives in those plots that had not been properly managed.

Effects of grazing can also depend on the individual plant species and its response to grazing. Plants that are adapted to extensive grazing (such as that done by cattle) will respond quicker and more effectively to grazing than native species that have not had to cope with intense grazing pressure in the past. An experiment done by Kimball and Schiffman showed that grazing increased the cover of some native species but did not decrease the cover of nonnative species. The species diversity of the native plants was able to respond to the grazing and increase diversity. The community would become denser than originally with the increased biodiversity. (However, this may have been simply variance in plots due to the fact that the native and nonnative compositions were of different species between the grazed and ungrazed plots).

Effects on Animals

Insects and Butterflies

Degree of grazing has a significant effect on the species richness and abundance of insects in grasslands. Land management in the form of grazing tends to decrease diversity with increased intensity. Kruess and Tscharntke attribute this difference to the increased height of grasses in the ungrazed areas. The study showed that the abundance and diversity of insects (such as butterfly adults, trap-nesting bees and wasps) were increased by increased grass height. However, other insects such as grasshoppers responded better to heterogeneity of the vegetation.

Vertebrates

Grazing can have varied effects on vertebrates. Different bird species react in different ways to changes in grazing intensity. Grazing has also been thought to decrease the abundance of vertebrates, such as the prairie dog and the desert tortoise. However, moderate grazing by cattle had no effect on the Texas tortoise.

Rabbits have been widely discussed due to their influences on land composition. *Bell and Watson* found that rabbits show grazing preference for different plant species. This preference can alter the composition of a plant community. In some cases, if the preference is for a non-native, invasive plant, rabbit grazing may benefit the community by reducing non-native abundance and creating room for the native plant species to fill. When rabbits graze in moderation they can create a more complex ecosystem, by creating more variable environments that will allow for more predator-competitor relationships between the various organisms. However, besides the effect on wild vegetation, rabbits destroy crops, compete with other herbivores, and can result in extreme ecological damage. Competition can be direct or indirect. The rabbits may specifically eat the competitions target food or it may inhibit the growth of grasses that other species eat. For example, rabbit grazing in the Netherlands inhibits tall grasses from becoming dominant. This in turn enhances the suitability of the pasture for brent goose. However, they may benefit predators that do better in open areas, because the rabbits reduce the amount of vegetation making it easier for those predators to spot their prey.

Finally, grazing has demonstrated use in clearing dry brush to reduce the fire hazard of drought-stricken areas.

Effect on Ephemeral Wetlands

Ephemeral wetlands degradation and loss of biodiversity had, at one point in time, been blamed on mismanaged grazing of both native and non-native ungulates and other grazers. A study done by Jaymee Marty of The Nature Conservancy examined the effects on the vernal pools formed in California when grazers were removed. The results of the short study showed that areas where grazers were removed had a lower diversity of native grasses, invertebrates and vertebrates in the pools, with an increase in non-native grass abundance and distribution in the area. The study also demonstrated reduced reproduction success of individual species in the area, such as the western spadefoot toad and California tiger salamander. Marty argues that this decrease is due to ecosystems adapting to historical changes in grazers and the effects they have. In other words, the historic ecosystem, theoretically, would have responded positively to the removal of cattle grazing, however, the system has adapted to the European introduced species and now may require them for maintained diversity. In another study performed by Pyke and Marty, measurements showed that on average, vernal ponds on grazed land pooled longer than ungrazed areas and soil was more resistant to water absorption in the grazed areas.

Targeted Grazing

A recent synonym or near-synonym for conservation grazing is "targeted grazing", a term introduced in a 2006 handbook in distinction to prescribed grazing, which the USDA National Resource Conservation Service was using to describe all managed grazing. Targeted grazing is often used in combination with other techniques such as burning, herbicide applications or land clearing. Targeted grazing can rival traditional herbicide and mechanical control methods for invasive plants from invasive forb to juniper trees, and has been used to reduce fine fuels in fire prone areas.

Principles

The most important skill for developing a targeted grazing program is patience and commitment. However, understanding livestock and plant responses to grazing are critical in developing a targeted grazing program. The program should have a clear statement of the kind of animal, timing and rate of grazing necessary to suppress troublesome plants and maintain a healthy landscape. The grazing application should 1) cause significant damage to the target plants 2) limit damage to desired vegetation and 3) be integrated with other control strategies. First, to cause significant damage to targeted plants requires understanding when the target plant is most susceptible to grazing damage and when they are most palatable to livestock. Target plant palatability depends on the grazing animals inherited and developed plant preferences (i.e. the shape of sheep and goat's mouths make them well suited for eating broad leaf weeds). Goats are also well designed for eating shrubs. Second, target plants often exist in a plant community with many desirable plants. The challenge is to select the correct animal, grazing time and grazing intensity to maximize the impact on the target plant while reducing it on the associated plant community. Finally, management objectives, target plant species, weather, topography, plant physiology, and associated plant communities are among the many variables that can determine treatment type and duration. Well-developed targeted grazing objectives and an adaptive management plan that takes into account other control strategies need to be in place.

Ecological Resilience

Ecological resilience, also called ecological robustness is the ability of an ecosystem to maintain its normal patterns of nutrient cycling and biomass production after being subjected to damage caused by an ecological disturbance. The term resilience is a term that is sometimes used interchangeably with robustness to describe the ability of a system to continue functioning amid and recover from a disturbance.

The resilience or robustness of ecological systems has been an important concept in ecology and natural history since the time of British naturalist Charles Darwin, who described the interdependencies between species as an "entangled bank" in his influential work On the Origin of Species. Since then, the concept has come to hold special importance in the areas of environmental conservation and management. Its significance to the well-being of humans and human societies has also been recognized. The loss of an ecosystem's ability to recover from a disturbance whether due to natural events such as hurricanes or volcanic eruptions or due to human influences such as overfishing and pollution endangers the benefits (e.g., food, clean water, and aesthetics) that humans derive from that ecosystem.

However, resilience is not always a positive feature of a system. For example, an ecosystem may be locked in an undesirable state, such as in the case of a eutrophic lake, where an overabundance of nutrients results in hypoxia (depleted oxygen levels), which can lead to the demise of desirable fish species and the proliferation of undesirable pests.

In 1955, Canadian-born American ecologist Robert MacArthur proposed a measure of community stability that was related to the complexity of an ecosystem's food web. He stated that ecosystem stability increased as the number of interactions (complexity) between the different species within the ecosystem also increased. His collaborator, Australian theoretical physicist Robert May, later showed that communities of species that were more diverse and more complex were actually less able to maintain an exact stable numerical balance among species. This seemingly counterintuitive idea occurs because resilience or robustness at the level of the ecosystem is actually enhanced by a lack of rigidity at the level of its individual components (i.e., the populations or species within the ecosystem). This elasticity means that ecosystem properties, such as changes in nutrient flow or the number of species, are more resilient due to changes in species composition. For example, the disappearance of the American chestnut (Castanea dentata) in many forests in eastern North America due to chestnut blight has been largely compensated for by the expansion of oak (Quercus) and hickory (Carya) species, although there are certainly commercial consequences of this replacement.

In 1973, Canadian ecologist C.S. Holling wrote a paper that focused on the dichotomy between a type of resilience inherent in an engineered device (that is, the stability that comes from a machine designed to operate within a narrow range of expected circumstances) and the resilience that emphasizes an ecosystem's persistence as a particular ecosystem type (e.g., a forest as opposed to grassland), the latter being affected by substantially more factors than the former. Holling recognized the importance of the qualities that allowed a forest to persist as a functioning forest rather than its ability to harbour particular species at fixed levels or to maintain an arbitrary level of primary production. Holling's seminal paper brought heightened attention to the resilience of ecological systems and influenced other disciplines, such as economics and sociology. It has resonated in particular with the perspectives of individuals such as American biophysicist and geographer Jared Diamond, who is known for his examination of the conditions under which human societies developed, thrived, and collapsed.

Resilience and the Development of Management Tools

Ecological resilience or robustness has also become central to conservation practices and ecosystem management, particularly as the latter has shifted its attention to the importance of ecosystem services. Such services include the provision of food, fuel, and natural products (e.g., substances for pharmaceutical development); the mediation of climate; the removal of toxic materials from environmental reservoirs; and the aesthetic enjoyment that humans derive from the natural world. Although many species retain importance within the framework of ecosystem services, much of the focus of conservation has moved from individual species to the maintenance of the ecosystem as a whole, especially its ability to retain its structure and rate of productivity.

Many lakes, for example, are managed to remain oligotrophic (relatively nutrient poor), with ample oxygen to support species such as lake trout, rather than managed to retain excess nutrients and algae. In addition, many terrestrial dryland ecosystems are managed to keep a richly vegetated area from undergoing desertification. Ecologists continue to look for ways to manage forests, such as those in Africa, to resist the transformation into a savanna through periods of extended drought or frequent wildfire episodes. Furthermore, in the ocean, where individual fish species have long been the subject of regulation, there is growing recognition of the need to expand efforts to manage large areas as integrated ecosystems.

Predicting the onset of disturbances such as eutrophication, desertification, and the collapse of fisheries has become an important component of ecosystem management. A greater emphasis on the identification of early-warning indicators, such as statistical fluctuations or correlations, has emerged. In particular, the ideas and techniques are being applied to medicine (such as in the onset of migraines or cardiac problems), research into climate change, and the operation of financial systems and markets. These indicators might serve as aids to management, much the same way that the detection of swarms of small earthquakes near a fault or an active volcano may portend the arrival of a larger seismic or eruptive event in the near future.

Equally important is the identification of the system's structural features that might impede the risk of systemic collapse or endow a system with the ability to recover from a disturbance. In ecological systems, ecologists might consider the diversity and heterogeneity among individual components (such as whole species, populations, or individual organisms) and landscape features within an ecosystem. Forest managers, for example, try to prevent the spread of wildfires throughout a forest by building firebreaks that follow changes in the landscape, such as those that separate one patch of trees from another. In addition, redundancy (niche overlap between species) and modularity (the interconnectedness of a system's components) are considered to be important factors that determine an ecosystem's resilience.

Phytoremediation

Phytoremediation, refers to the technologies that use living plants to clean up soil, air, and water contaminated with hazardous contaminants. It is defined as "the use of green plants and the associated microorganisms, along with proper soil amendments and agronomic techniques to either contain, remove or render toxic environmental contaminants harmless".

Phytoremediation is a cost-effective plant-based approach of remediation that takes advantage of the ability of plants to concentrate elements and compounds from the environment and to metabolize various molecules in their tissues. It refers to the natural ability of certain plants called hyperaccumulators to bioaccumulate, degrade, or render harmless contaminants in soils, water, or air. Toxic heavy metals and organic pollutants are the major targets for phytoremediation. Knowledge of the physiological and molecular mechanisms of phytoremediation began to emerge in recent years together with biological and engineering strategies designed to optimize and improve phytoremediation. In addition, several field trials confirmed the feasibility of using plants for environmental cleanup.

Application

Phytoremediation may be applied wherever the soil or static water environment has become polluted or is suffering ongoing chronic pollution. Examples where phytoremediation has been used successfully include the restoration of abandoned metal mine workings, and sites where polychlorinated biphenyls have been dumped during manufacture and mitigation of ongoing coal mine discharges reducing the impact of contaminants in soils, water, or air. Contaminants such as metals, pesticides, solvents, explosives, and crude oil and its derivatives, have been mitigated in phytoremediation projects worldwide. Many plants such as mustard plants, alpine pennycress, hemp, and pigweed have proven to be successful at hyperaccumulating contaminants at toxic waste sites.

Not all plants are able to accumulate heavy metals or organics pollutants due to differences in the physiology of the plant. Even cultivars within the same species have varying abilities to accumulate pollutants.

Over the past 20 years, this technology has become increasingly popular and has been employed at sites with soils contaminated with lead, uranium, and arsenic. While it has the advantage that environmental concerns may be treated in situ, one major disadvantage of phytoremediation is that it requires a long-term commitment, as the process is dependent on a plant's ability to grow and thrive in an environment that is not ideal for normal plant growth.

Advantages and Limitations

- Advantages:
    - The cost of the phytoremediation is lower than that of traditional processes both in situ and ex situ.
    - The plants can be easily monitored.
    - The possibility of the recovery and re-use of valuable metals (by companies specializing in "phyto mining").
    - It is potentially the least harmful method because it uses naturally occurring organisms and preserves the environment in a more natural state.
    - It preserves the topsoil, maintaining the fertility of the soil.
    - Increase soil health, yield, and plant phytochemicals.
    - The use of plants also reduces erosion and metal leaching in the soil.
- Limitations:
    - Phytoremediation is limited to the surface area and depth occupied by the roots.
    - Slow growth and low biomass require a long-term commitment.
    - With plant-based systems of remediation, it is not possible to completely prevent the leaching of contaminants into the groundwater (without the

complete removal of the contaminated ground, which in itself does not resolve the problem of contamination).

- The survival of the plants is affected by the toxicity of the contaminated land and the general condition of the soil.
- Bio-accumulation of contaminants, especially metals, into the plants which then pass into the food chain, from primary level consumers upwards or requires the safe disposal of the affected plant material.
- When taking up heavy metals, sometimes the metal is bound to the soil organic matter, which makes it unavailable for the plant to extract.

Processes

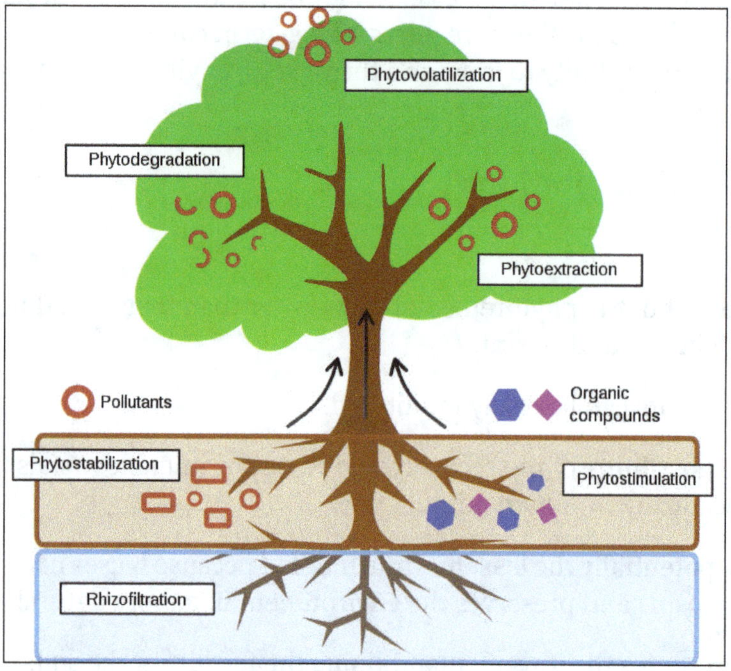

Phytoremediation process.

A range of processes mediated by plants or algae are useful in treating environmental problems.

Phytoextraction

Phytoextraction (or phytoaccumulation or phytosequestration) uses plants or algae to remove contaminants from soil or water into harvestable plant biomass. The roots take up substances from the soil or water and concentrate it above ground in the plant biomass Organisms that can uptake extremely high amounts of contaminants from the soil are called hyperaccumulators. Phytoextraction can also be performed by plants (e.g. Populus and Salix) that take up lower levels of pollutants, but due to their high growth rate and biomass production, may remove a considerable amount of contaminants from the soil. Phytoextraction has been growing rapidly in popularity worldwide for the last twenty years or so. Typically, phytoextraction is used for heavy metals or other inorganics. At the time of disposal, contaminants are typically concentrated in the much smaller volume of the plant matter than in the initially contaminated soil or sediment. After harvest, a lower level of the contaminant will remain in the soil, so the growth/harvest cycle must usually be repeated through several crops to achieve a significant cleanup. After the process, the cleaned soil can support other vegetation. Mining of these extracted metals through phytomining, is also being experimented with as a way of recovering the material. Hyperaccumulators are plants that can naturally take up the contaminants in soil unassisted. In many cases these are metallophyte plants that can tolerate and incorporate high levels of toxic metals. Induced or assisted phytoextraction is a process where a conditioning fluid containing a chelator or another agent is added to soil to increase metal solubility or mobilization so that the plants can absorb them more easily. While this leads to increased metal uptake by plants, it can also lead to large amounts of available metals in the soil beyond what the plants are able to translocate, causing potential leaching into the subsoil or groundwater.

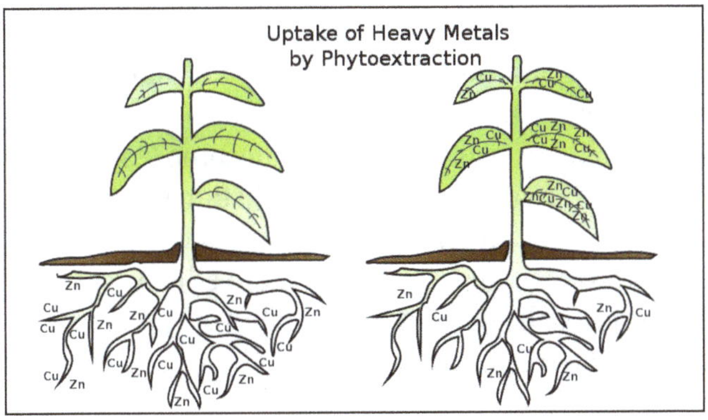

Some heavy metals such as copper and zinc are removed from the soil by moving up into the plant roots.

Examples of plants that are known to accumulate the following contaminants:

- Arsenic, using the sunflower (Helianthus annuus), or the Chinese Brake fern (Pteris vittata).
- Cadmium, using willow (Salix viminalis): In 1999, one research experiment performed by Maria Greger and Tommy Landberg suggested willow has a significant potential as a phytoextractor of cadmium (Cd), zinc (Zn), and copper (Cu), as willow has some specific characteristics like high transport capacity of heavy metals from root to shoot and huge amount of biomass production; can be used also for production of bio energy in the biomass energy power plant.
- Cadmium and zinc, using alpine pennycress (Thlaspi caerulescens), a hyperaccumulator of these metals at levels that would be toxic to many plants. On the other hand, the presence of copper seems to impair its growth.

- Lead, using Indian mustard (Brassica juncea), ragweed (Ambrosia artemisiifolia), hemp dogbane (Apocynum cannabinum), or poplar trees, which sequester lead in their biomass.
- Salt-tolerant (moderately halophytic) barley and/or sugar beets are commonly used for the extraction of sodium chloride (common salt) to reclaim fields that were previously flooded by sea water.
- Caesium-137 and strontium-90 were removed from a pond using sunflowers after the Chernobyl accident.
- Mercury, selenium and organic pollutants such as polychlorinated biphenyls (PCBs) have been removed from soils by transgenic plants containing genes for bacterial enzymes.

Phytostabilization

Phytostabilization reduces the mobility of substances in the environment, for example, by limiting the leaching of substances from the soil. It focuses on the long term stabilization and containment of the pollutant. The plant immobilizes the pollutants by binding them to soil particles making them less available for plant or human uptake. Unlike phytoextraction, phytostabilization focuses mainly on sequestering pollutants in soil near the roots but not in plant tissues. Pollutants become less bioavailable, resulting in reduced exposure. The plants can also excrete a substance that produces a chemical reaction, converting the heavy metal pollutant into a less toxic form. Stabilization results in reduced erosion, runoff, leaching, in addition to reducing the bioavailability of the contaminant. An example application of phytostabilization is using a vegetative cap to stabilize and contain mine tailings.

Phytodegradation

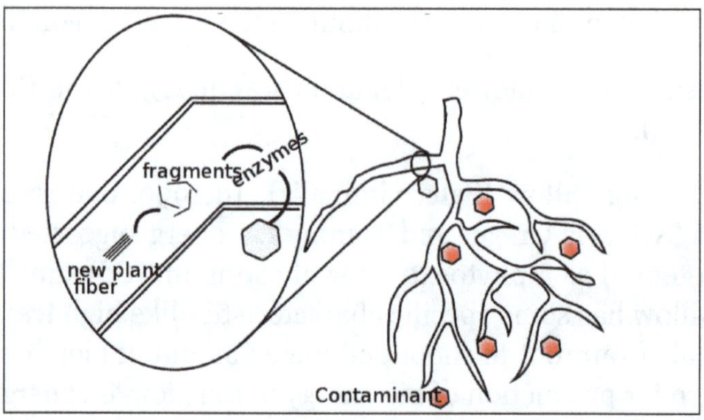

The roots secrete enzymes that degrade (breakdown) organic pollutants in the soil.

Phytodegradation (also called phytotransformation) uses plants or microorganisms to degrade organic pollutants in the soil or within the body of the plant. The organic compounds are broken down by enzymes that the plant roots secrete and these molecules are then taken up by the plant and released through transpiration. This process works best with organic contaminants like herbicides, trichloroethylene, and methyl tert-butyl ether.

Phytotransformation results in the chemical modification of environmental substances as a direct result of plant metabolism, often resulting in their inactivation, degradation (phytodegradation), or immobilization (phytostabilization). In the case of organic pollutants, such as pesticides, explosives, solvents, industrial chemicals, and other xenobiotic substances, certain plants, such as Cannas, render these substances non-toxic by their metabolism. In other cases, microorganisms living in association with plant roots may metabolize these substances in soil or water. These complex and recalcitrant compounds cannot be broken down to basic molecules (water, carbon-dioxide, etc.) by plant molecules, and, hence, the term phytotransformation represents a change in chemical structure without complete breakdown of the compound. The term "Green Liver" is used to describe phytotransformation, as plants behave analogously to the human liver when dealing with these xenobiotic compounds (foreign compound/pollutant). After uptake of the xenobiotics, plant enzymes increase the polarity of the xenobiotics by adding functional groups such as hydroxyl groups (-OH).

This is known as Phase I metabolism, similar to the way that the human liver increases the polarity of drugs and foreign compounds (drug metabolism). Whereas in the human liver enzymes such as cytochrome P450s are responsible for the initial reactions, in plants enzymes such as peroxidases, phenoloxidases, esterases and nitroreductases carry out the same role.

In the second stage of phytotransformation, known as Phase II metabolism, plant biomolecules such as glucose and amino acids are added to the polarized xenobiotic to further increase the polarity (known as conjugation). This is again similar to the processes occurring in the human liver where glucuronidation (addition of glucose molecules by the UGT class of enzymes, e.g. UGT1A1) and glutathione addition reactions occur on reactive centres of the xenobiotic.

Phase I and II reactions serve to increase the polarity and reduce the toxicity of the compounds, although many exceptions to the rule are seen. The increased polarity also allows for easy transport of the xenobiotic along aqueous channels.

In the final stage of phytotransformation (Phase III metabolism), a sequestration of the xenobiotic occurs within the plant. The xenobiotics polymerize in a lignin-like manner and develop a complex structure that is sequestered in the plant. This ensures that the xenobiotic is safely stored, and does not affect the functioning of the plant. However, preliminary studies have shown that these plants can be toxic to small animals (such as snails), and, hence, plants involved in phytotransformation may need to be maintained in a closed enclosure.

Hence, the plants reduce toxicity (with exceptions) and sequester the xenobiotics in phytotransformation. Trinitrotoluene phytotransformation has been extensively researched and a transformation pathway has been proposed.

Phytostimulation

Phytostimulation (or rhizodegradation) is the enhancement of soil microbial activity for the degradation of organic contaminants, typically by organisms that associate with roots. This process occurs within the rhizosphere, which is the layer of soil that surrounds the roots. Plants release carbohydrates and acids that stimulate microorganism activity which results in the biodegradation of the organic contaminants. This means that the microorganisms are able to digest and break down the toxic substances into harmless form. Phytostimulation has been shown to be effective in degrading petroleum hydrocarbons, PCBs, and PAHs. Phytostimulation can also involve aquatic plants supporting active populations of microbial degraders, as in the stimulation of atrazine degradation by hornwort.

Phytovolatilization

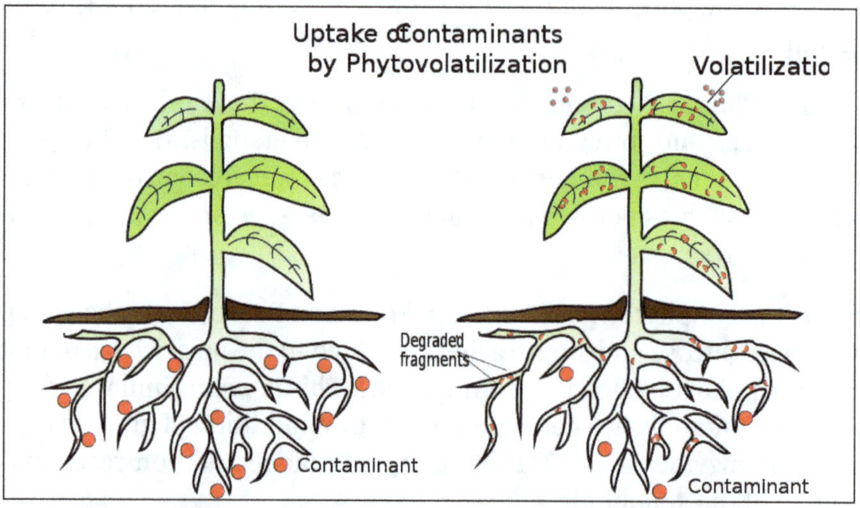

Contaminates are then broken down and the fragments are then subsequently transformed and volatilized into the atmosphere.

Phytovolatilization is the removal of substances from soil or water with release into the air, sometimes as a result of phytotransformation to more volatile and/or less polluting substances. In this process, contaminants are taken up by the plant and through transpiration, evaporate into the atmosphere. This is the most studied form of phytovolatilization, where volatilization occurs at the stem and leaves of the plant, however indirect phytovolatilization occurs when contaminants are volatilized from the root zone. Selenium (Se) and Mercury (Hg) are often removed from soil through phytovolatilization. Poplar trees are one of the most successful plants for removing VOCs through this process due to its high transpiration rate.

Rhizofiltration

Rhizofiltration is a process that filters water through a mass of roots to remove toxic substances or excess nutrients. The pollutants remain absorbed in or adsorbed to the roots. This process is often used to clean up contaminated groundwater through planting directly in the contaminated site or through removing the contaminated water and providing it to these plants in an off-site location. In either case though, typically plants are first grown in a greenhouse under precise conditions.

Biological Hydraulic Containment

Biological hydraulic containment occurs when some plants, like poplars, draw water upwards through the soil into the roots and out through the plant, which decreases the movement of soluble contaminants downwards, deeper into the site and into the groundwater.

Phytodesalination

Phytodesalination uses halophytes (plants adapted to saline soil) to extract salt from the soil to improve its fertility.

Role of Genetics

Breeding programs and genetic engineering are powerful methods for enhancing natural phytoremediation capabilities, or for introducing new capabilities into plants. Genes for phytoremediation may originate from a micro-organism or may be transferred from one plant to another variety better adapted to the environmental conditions at the cleanup site. For example, genes encoding a nitroreductase from a bacterium were inserted into tobacco and showed faster removal of TNT and enhanced resistance to the toxic effects of TNT. Researchers have also discovered a mechanism in plants that allows them to grow even when the pollution concentration in the soil is lethal for non-treated plants. Some natural, biodegradable compounds, such as exogenous polyamines, allow the plants to tolerate concentrations of pollutants 500 times higher than untreated plants, and to absorb more pollutants.

Hyperaccumulators and Biotic Interactions

A plant is said to be a hyperaccumulator if it can concentrate the pollutants in a minimum percentage which varies according to the pollutant involved (for example: more than 1000 mg/kg of dry weight for nickel, copper, cobalt, chromium or lead; or more than 10,000 mg/kg for zinc or manganese). This capacity for accumulation is due to hypertolerance, or phytotolerance: the result of adaptative evolution from the plants to hostile environments through many generations. A number of interactions may be affected by metal hyperaccumulation, including protection, interferences with neighbour plants of different species, mutualism (including mycorrhizae, pollen and seed dispersal), commensalism, and biofilm.

Phytoscreening

As plants are able to translocate and accumulate particular types of contaminants, plants can be used as biosensors of subsurface contamination, thereby allowing investigators to quickly delineate contaminant plumes. Chlorinated solvents, such as trichloroethylene, have been observed in tree trunks at concentrations related to groundwater concentrations. To ease field implementation of phytoscreening, standard methods have been developed to extract a section of the tree trunk for later laboratory analysis, often by using an increment borer. Phytoscreening may lead to more optimized site investigations and reduce contaminated site cleanup costs.

Plant Litter

Leaf litter, mainly White Beech, Gmelina leichhardtii, from Black Bulga State Conservation Area, NSW, Australia.

Litterfall, plant litter, leaf litter, tree litter, soil litter, or duff, is dead plant material (such as leaves, bark, needles, twigs, and cladodes) that have fallen to the ground. This detritus or dead organic material and its constituent nutrients are added to the top layer of soil, commonly known as the litter layer or O horizon ("O" for "organic"). Litter has occupied the attention of ecologists at length for the reasons that it is an instrumental factor in ecosystem dynamics, it is indicative of ecological productivity, and may be useful in predicting regional nutrient cycling and soil fertility.

Characteristics and Variability

Plant litter, mainly western hemlock, Tsuga heterophylla, in Mount Baker-Snoqualmie National Forest, Washington, United States.

Litterfall is characterized as fresh, undecomposed, and easily recognizable (by species and type) plant debris. This can be anything from leaves, cones, needles, twigs, bark, seeds/nuts, logs, or reproductive organs (e.g. the stamen of flowering plants). Items larger than 2 cm diameter are referred to as coarse litter, while anything smaller is referred to as fine litter or litter. The type of litterfall is most directly affected by ecosystem type. For example, leaf tissues account for about 70 percent of litterfall in forests, but woody litter tends to increase with forest age. In grasslands, there is very little aboveground perennial tissue so the annual litterfall is very low and quite nearly equal to the net primary production.

In soil science, soil litter is classified in three layers, which form on the surface of the O Horizon. These are the L, F, and H layers:

- L: Organic horizon characterized by relatively undecomposed plant material.
- F: Organic horizon found beneath L characterized by accumulation of partly decomposed organic matter.
- H: Organic horizon below F characterized by accumulation of fully decomposed organic matter mostly indiscernible.

The litter layer is quite variable in its thickness, decomposition rate and nutrient content and is affected in part by seasonality, plant species, climate, soil fertility, elevation, and latitude. The most extreme variability of litterfall is seen as a function of seasonality; each individual species of plant has seasonal losses of certain parts of its body, which can be determined by the collection and classification of plant litterfall throughout the year, and in turn affects the thickness of the litter layer. In tropical environments, the largest amount of debris falls in the latter part of dry seasons and early during wet season. As a result of this variability due to seasons, the decomposition rate for any given area will also be variable.

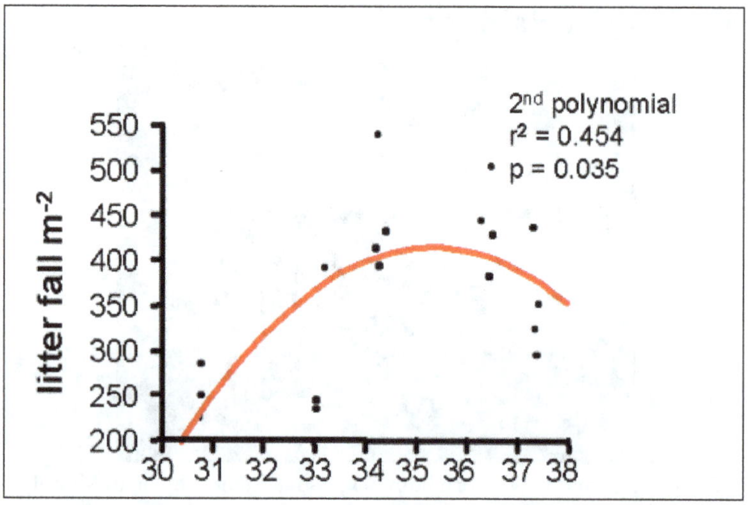

Litter fall in the North American Baldcypress Swamp Network, Illinois to Louisiana.

Latitude also has a strong effect on litterfall rates and thickness. Specifically, litterfall declines with increasing latitude. In tropical rainforests, there is a thin litter layer due to the rapid decomposition, while in boreal forests, the rate of decomposition is slower and leads to the accumulation of a thick litter layer, also known as a mor. Net primary production works inversely to this trend, suggesting that the accumulation of organic matter is mainly a result of decomposition rate.

Surface detritus facilitates the capture and infiltration of rainwater into lower soil layers. Soil litter protects soil aggregates from raindrop impact, preventing the release of clay and silt particles from plugging soil pores. Releasing clay and silt particles reduces the capacity for soil to absorb water and increases cross surface flow, accelerating soil erosion. In addition soil litter reduces wind erosion by preventing soil from losing moisture and providing cover preventing soil transportation.

Organic matter accumulation also helps protect soils from wildfire damage. Soil litter can be completely removed depending on intensity and severity of wildfires and season. Regions with high frequency wildfires have reduced vegetation density and reduced soil litter accumulation. Climate also influences the depth of plant litter. Typically humid tropical and sub-tropical climates have reduced organic matter layers and horizons due to year round decomposition and high vegetation density and growth. In temperate and cold climates, litter tends to accumulate and decompose slower due to a shorter growing season.

Net Primary Productivity

Net primary production and litterfall are intimately connected. In every terrestrial ecosystem, the largest fraction of all net primary production is lost to herbivores and litterfall. Due to their interconnectedness, global patterns of litterfall are similar to global patterns of net primary productivity.

Habitat and Food

Litter provides habitat for a variety of organisms.

Plants

Common wood sorrel (Oxalis acetosella) in Ivanovo Oblast, Russia.

Certain plants are specially adapted for germinating and thriving in the litter layers. For example, bluebell (Hyacinthoides non-scripta) shoots puncture the layer to emerge in spring. Some plants with rhizomes, such as common wood sorrel (Oxalis acetosella) do well in this habitat.

Detritivores and other Decomposers

Fungi in the forest floor (Marselisborg Forests in Denmark).

Many organisms that live on the forest floor are decomposers, such as fungi. Organisms whose diet consists of plant detritus, such as earthworms, are termed detritivores. The community of decomposers in the litter layer also includes bacteria, amoeba, nematodes, rotifer, tardigrades, springtails, cryptostigmata, potworms, insect larvae, mollusks, oribatid mites, woodlice, and millipedes. Their consumption of the litterfall results in the breakdown of simple carbon compounds into carbon dioxide (CO_2) and water (H_2O), and releases inorganic ions (like nitrogen and phosphorus) into the soil where the surrounding plants can then reabsorb the nutrients that were shed as litterfall. In this way, litterfall becomes an important part of the nutrient cycle that sustains forest environments.

A skink, Eutropis multifasciata, in leaf litter in Sabah, Malaysia.

As litter decomposes, nutrients are released into the environment. The portion of the litter that is not readily decomposable is known as humus. Litter aids in soil moisture retention by cooling the ground surface and holding moisture in decaying organic matter. The flora and fauna working to decompose soil litter also aid in soil respiration. A litter layer of decomposing biomass provides a continuous energy source for macro- and micro-organisms.

Larger Animals

Numerous reptiles, amphibians, birds, and even some mammals rely on litter for shelter and forage. Amphibians such as salamanders and caecilians inhabit the damp microclimate underneath fallen leaves for part or all of their life cycle. This makes them difficult to observe. A BBC film crew captured footage of a female caecilian with young for the first time in a documentary that aired in 2008. Some species of birds, such as the ovenbird of eastern North America for example, require leaf litter for both foraging and material for nests. Sometimes litterfall even provides energy to much larger mammals, such as in boreal forests where lichen litterfall is one of the main constituents of wintering deer and elk diets.

Nutrient Cycle

During leaf senescence, a portion of the plant's nutrients are reabsorbed from the leaves. The nutrient concentrations in litterfall differ from the nutrient concentrations in the mature foliage by the reabsorption of constituents during leaf senescence. Plants that grow in areas with low nutrient availability tend to produce litter with low nutrient concentrations, as a larger proportion of the available nutrients is reabsorbed.

After senescence, the nutrient-enriched leaves become litterfall and settle on the soil below.

Litterfall is the dominant pathway for nutrient return to the soil, especially for nitrogen (N) and phosphorus (P). The accumulation of these nutrients in the top layer of soil is known as soil immobilization. Once the litterfall has settled, decomposition of the litter layer, accomplished through the leaching of nutrients by rainfall and throughfall and by the efforts of detritivores, releases the breakdown products into the soil below and therefore contributes to the cation exchange capacity of the soil. This holds especially true for highly weathered tropical soils.

Leaching is the process by which cations such as iron (Fe) and aluminum (Al), as well as organic matter are removed from the litterfall and transported downward into the soil below. This process is known as podzolization and is particularly intense in boreal and cool temperate forests that are mainly constituted by coniferous pines whose litterfall is rich in phenolic compounds and fulvic acid.

By the process of biological decomposition by microfauna, bacteria, and fungi, CO_2 and H_2O, nutrient elements, and a decomposition-resistant organic substance called humus are released. Humus composes the bulk of organic matter in the lower soil profile.

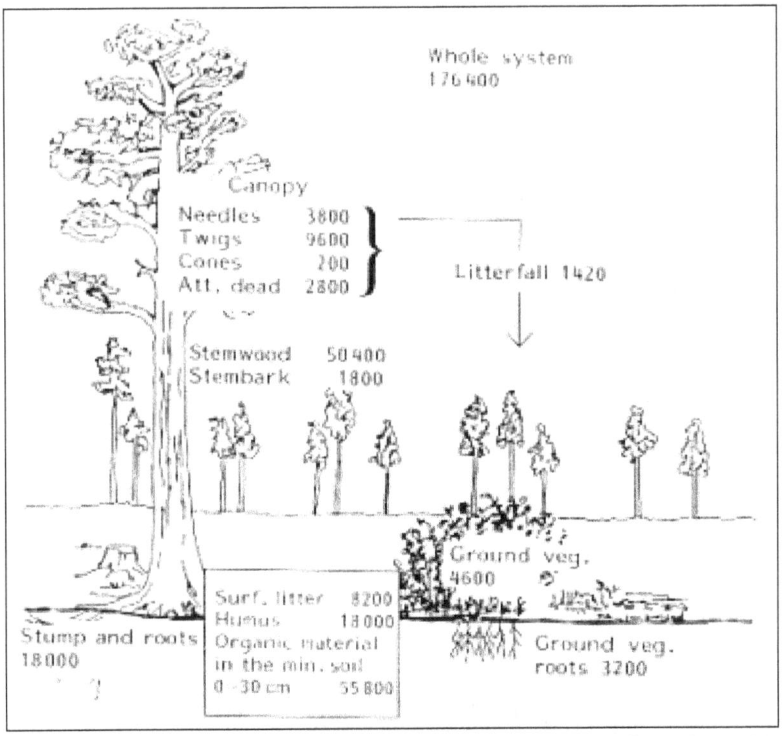

A budget for organic matter in a mature (120-year-old) Scots pine monoculture (SWECON site). Based on data from Andersson. Units are in kg of organic matter per ha. Att. -attached; Surf. -surface; min. -mineral; and veg. -vegetation.

The decline of nutrient ratios is also a function of decomposition of litterfall (i.e. as litterfall decomposes, more nutrients enter the soil below and the litter will have a lower nutrient ratio). Litterfall containing high nutrient concentrations will decompose more rapidly and asymptote as those nutrients decrease. Knowing this, ecologists have been able to use nutrient concentrations as measured by remote sensing as an index of a potential rate of decomposition for any given area. Globally, data from various forest ecosystems shows an inverse relationship in the decline in nutrient ratios to the apparent nutrition availability of the forest.

Once nutrients have re-entered the soil, the plants can then reabsorb them through their roots. Therefore, nutrient reabsorption during senescence presents an opportunity for a plant's future net primary production use. A relationship between nutrient stores can also be defined as:

> annual storage of nutrients in plant tissues + replacement of losses from litterfall and leaching = the amount of uptake in an ecosystem.

Collection and Analysis

The main objectives of litterfall sampling and analysis are to quantify litterfall production and chemical composition over time in order to assess the variation in litterfall quantities, and hence its role in nutrient cycling across an environmental gradient of climate (moisture and temperature) and soil conditions.

Ecologists employ a simple approach to the collection of litterfall, most of which centers around one piece of equipment, known as a litterbag. A litterbag is simply any type of container that can be set out in any given area for a specified amount of time to collect the plant litter that falls from the canopy above.

Litterfall and throughfall collectors at beech stand in Thetford, East Anglia.

Litterbags are generally set in random locations within a given area and marked with GPS or local coordinates, and then monitored on a specific time interval. Once the samples have been collected, they are usually classified on type, size and species (if possible) and recorded on a spreadsheet. When measuring bulk litterfall for an area, ecologists will weigh the dry contents of the litterbag. By this method litterfall flux can be defined as:

litterfall (kg m^{-2} yr^{-1}) = total litter mass (kg) / litterbag area (m^2)

The litterbag may also be used to study decomposition of the litter layer. By confining fresh litter in the mesh bags and placing them on the ground, an ecologist can monitor and collect the decay measurements of that litter. An exponential decay pattern has been produced by this type of experiment: $\frac{X}{X_o} = e^{-k}$, where X_o is the initial leaf litter and k is a constant fraction of detrital mass.

The mass-balance approach is also utilized in these experiments and suggests that the decomposition for a given amount of time should equal the input of litterfall for that same amount of time.

litterfall = k(detrital mass)

For study various groups from edaphic fauna you need a different mesh sizes in the litterbags.

Issues

Change due to Invasive Earthworms

In some regions of glaciated North America, earthworms have been introduced where they are not native. Ecologists call this process succession. Non-native earthworms have led to environmental changes by accelerating the rate of decomposition of litter. These changes are being studied, but may have negative impacts on some inhabitants such as salamanders.

Forest Litter Raking

This thin, delicate layer of organic material can be easily affected by humans. For instance, forest litter raking as a replacement for straw in husbandry is an old non-timber practice in forest management that has been widespread in Europe since the seventeenth century. In 1853, an estimated 50 Tg of dry litter per year was raked in European forests, when the practice reached its peak. This human disturbance, if not combined with other degradation factors, could promote podzolisation; if managed properly (for example, by burying litter removed after its use in animal husbandry), even the repeated removal of forest biomass may not have negative effects on pedogenesis.

Stream Restoration

Robinson Creek in Boonville, California had highly eroded stream banks prior to initiation of a stream restoration project.

Stream restoration or river restoration, also sometimes referred to as river reclamation, describes work conducted to improve the environmental health of a river or stream, in support of biodiversity, recreation, flood management and/or landscape development. Stream restoration approaches can be divided into two broad categories: form-based restoration, which relies on physical interventions in the stream to improve its conditions; and process-based restoration, which advocates the restoration of hydrological and geomorphological processes (such as sediment transport or connectivity between the channel and the floodplain) to ensure the stream's resilience and ecological health. Form-based restoration techniques include deflectors; cross-vanes; weirs, step-pools and other grade control structures; engineered log jams; bank stabilization methods and other channel reconfiguration efforts. These induce immediate change in the stream, but sometimes fail to achieve the desired effects if degradation originates at a wider scale. Process-based restoration includes restoring lateral or longitudinal connectivity of water and sediment fluxes and limiting interventions within in a corridor defined based on the stream's hydrology and geomorphology. The beneficial effects of process-based restoration projects may sometimes take time to be felt since changes in the stream will occur at a pace that depends on the stream dynamics.

Despite the significant number of stream restoration projects worldwide, the effectiveness of stream restoration remains poorly quantified, partly due to insufficient monitoring. However, in response to growing environmental awareness, stream restoration requirements are increasingly adopted in legislation in various parts of the world.

Stream restoration or river restoration, sometimes called river reclamation in the United Kingdom, describes a set of activities that aim to improve the environmental health of a river or stream. These activities aim to restore the river system to its original state or to a reference state, in support of biodiversity, recreation, flood management and/ or landscape development. Stream restoration is generally associated to environmental restoration and ecological restoration. In that sense, stream restoration differs from:

- River engineering, a term which typically refers to physical alterations of a water body, for purposes that include navigation, flood control or water supply diversion and are not necessarily related to ecological restoration.
- Waterway restoration, a term used in the United Kingdom describing alterations to a canal or river to improve navigability and related recreational amenities.

Improved stream health may be indicated by expanded habitat for diverse species (e.g. fish, aquatic insects, other wildlife) and reduced stream bank erosion, although bank erosion is increasingly generally recognized as contributing to the ecological health of streams. Enhancements may also include improved water quality (i.e. reduction of pollutant levels and increase of dissolved oxygen levels) and achieving a self-sustaining, resilient stream system that does not require periodic human intervention, such as dredging or construction of flood or erosion control structures. Stream restoration projects can also yield increased property values in adjacent areas.

In the past decades, stream restoration has emerged as a significant discipline in the field of water resources management, due to the degradation of many aquatic and riparian ecosystems related to human activities. In the USA alone, it was estimated in the early 2000s that more than one billion US dollars were spent each year to restore rivers and that close to 40,000 restoration projects had been conducted in the continental part of the country.

Restoration Approaches and Techniques

Robinson Creek restoration project included re-shaping of stream bank slopes, addition of live willow and large rock baffles, removal of invasive species and revegetation with indigenous species.

Stream restoration activities may range from the simple improvement or removal of a structure that inhibits natural stream functions (e.g. repairing or replacing a culvert, or removing barriers to fish passage such as weirs), to the stabilization of stream banks, or other interventions such as riparian zone restoration or the installation of stormwater management facilities like constructed wetlands. The use of recycled water to augment stream flows that have been depleted as a result of human activities can also be considered a form of stream restoration. When present, navigation locks have a potential to be operated as vertical slot fishways to restore fish passage to some extent for a wide range of biota, including poor swimmers.

Stream restoration projects normally begin with an assessment of the stream system, including climatic data, geology, watershed hydrology, stream hydraulics, sediment transport patterns, channel geometry, historical channel mobility and flood records. Numerous systems exist to classify streams according to their geomorphology. This preliminary assessment helps to understand the stream dynamics and determining the cause of the observed degradation to be addressed; it can also be used to determine the target state for the intended restoration work, especially since the "natural" or undisturbed state is sometimes no longer achievable due to various constraints.

Two broad approaches to stream restoration have been defined in the past decades: form-based restoration and process-based restoration. Whereas the former focuses on the restoration of structural features and/or patterns considered to be characteristic of the target stream system, the latter is based on the restoration of hydrological and geomorphological processes (such as sediment transport or connectivity between the channel and the floodplain) to ensure the stream's resilience and ecological health.

Form-based Restoration

Form-based stream restoration promotes the modification of the stream channel to improve the stream conditions. Targeted outcomes can include improved water quality, enhanced fish habitat and abundance, as well as increased bank and channel stability. This approach is widely used worldwide, and is supported by various government agencies, including the United States Environmental Protection Agency (US EPA).

Form-based restoration projects can be carried out at various scales, including the reach scale. They can include measures such as the installation of in-stream structures, bank stabilization and more significant channel reconfiguration efforts. Reconfiguration work may focus on channel shape (in terms of sinuosity and meander characteristics), cross-section or channel profile (slope along the channel bed). These alterations affect the dissipation of energy through the channel, which impacts flow velocity and turbulence, water surface elevations, sediment transport and scour, among other characteristics.

Installation of In-stream Structures

Deflectors

Deflectors are generally wooden or rock structures installed at the bank toe and extending towards the center of the stream, in order to concentrate stream flow away from the banks. They can limit bank erosion and generate varying flow conditions in terms of depth and velocity, which can positively impact fish habitat.

Cross-vanes and Related Structures

Cross-vanes are "U"-shaped structures made of boulders or logs, built across the channel to concentrate stream flow in the center of the channel and thereby reduce bank erosion. They do not impact channel capacity and provides other benefits such as improved habitat for aquatic species. Similar structures used to dissipate stream energy include the W-weirs and J-Hook vanes.

Cross-vanes.

Weirs, Step Pools and Grade Control Structures

Boulder step pools installed in Rock Creek, Washington, D.C. The pools raise the water level and allow fish to swim over a partially-submerged sewer pipe which crosses the creek.

These structures, which can be built with rocks or wood (logs or woody debris), gradually lower the elevation of the stream and dissipate flow energy, thereby reducing flow velocity. They can help limiting bed degradation. They generate water accumulation upstream from them and fast flowing conditions downstream from them, which can improve fish habitat. However, they can limit fish passage if they are too high.

Engineered Log Jams

An emerging stream restoration technique is the installation of engineered log jams. Because of channelization and removal of beaver dams and woody debris, many streams lack the hydraulic complexity that is necessary to maintain bank stabilization and healthy aquatic habitats. Reintroduction of large woody debris into streams is a method that is being experimented in streams such as Lagunitas Creek in Marin County, California and Thornton Creek, in Seattle, Washington. Log jams add diversity to the water flow by creating riffles, pools, and temperature variations. Large wood pieces, both living and dead, play an important role in the long-term stability of engineered log jams. However, individual pieces of wood in log jams are rarely stable over long time periods and are naturally transported downstream, where they can get trapped in further log jams, other stream features or human infrastructures, which can generate nuisances for human use.

Bank Stabilization

Bank stabilization is a common objective for stream restoration projects, although bank erosion is generally viewed as favorable for the sustainability and diversity of aquatic and riparian habitats. This technique may be employed where a stream reach is highly confined, or where infrastructure is threatened.

Bank stabilization is achieved through the installation of riprap, gabions or through the use of revegetation and/or bioengineering methods, which relies on the use of live plants to build bank stabilizing structures. As new plants sprout from the live branches, the roots anchor the soil and prevent erosion. This makes bioengineering structures more natural and more adaptable to evolving conditions than "hard" engineering structures. Bioengineering structures include fascines, brush mattresses, brush layer, vegetated geogrids, etc.

Other Channel Reconfiguration Techniques

Channel reconfiguration involves the physical modification of the stream. Depending on the scale of the projects, the channel's cross-section can be modified, and meanders can be constructed through earthworks to achieve the target morphology. In the USA, such work is frequently based on the Natural Channel Design (NCD), a method developed in the 1990s. This method involves a classification of the stream to be restored based on parameters such as channel pattern and geometry, topography, slope, and bed material. This classification is followed by a design phase based on the NCD method, which includes 8 phases and 40 steps. The method relies on the construction of the desired morphology, and its stabilization with natural materials such as boulders and vegetation to limit erosion and channel mobility.

Criticisms to Form-based Restoration

Despite its popularity, form-based restoration has been criticized by the scientific community. Common criticisms are that the scale at which form-based restoration is often much smaller than the spatial and temporal scales of the processes that cause the observed problems and that the target state is frequently influenced by the social conception of what a stream should look like and does not necessarily take into account the stream's geomorphological context (e.g. meandering rivers tend to be viewed as more "natural" and more beautiful, whereas local conditions sometimes favour other patterns such as braided rivers). Numerous criticisms have also been directed at the NCD method by fluvial geomorphologists, which claim that the method is a "cookbook" approach sometimes used by practitioners that do not have sufficient knowledge of fluvial geomorphology, resulting in project failures. Another criticism is the importance given to channel stability in the NCD method (and with some other form-based restoration methods), which can limit the streams' alluvial dynamic and adaptability to evolving conditions.

Process-based Restoration

Contrary to form-based restoration, which consists of improving a stream's conditions by modifying its structure, process-based restoration focuses on restoring the hydrological and geomorphological processes (or functions) that contribute to the stream's alluvial and ecological dynamics. This type of stream restoration has gained in popularity since the mid-1990s, as a more ecosystem-centered approach. Process-based restoration includes restoring lateral connectivity (between the stream and its floodplain), longitudinal connectivity (along the stream) and water and/or sediment fluxes, which might be impacted by hydro-power dams, grade control structures, erosion control structures and flood protection structures. In general, process-based restoration aims to maximize the resilience of the system and minimize maintenance requirements. In some instances, form-based restoration methods might be coupled with process-based restoration to restore key structures and achieve quicker results while waiting for restored processes to ensure adequate conditions in the long term.

Improving Connectivity

The connectivity of streams to their adjacent floodplain and along their entire length plays an important role in the equilibrium of the river system. Streams are shaped by the water and sediment fluxes from their watershed, and any alteration of these fluxes (either in quantity, intensity or timing) will result in changes in equilibrium planform and cross-sectional geometry, as well as modifications of the aquatic and riparian ecosystem. Removal or modification of levees can allow a better connection between streams and their floodplain. Similarly, removing dams and grade control structures can restore water and sediment fluxes and result in more diversified habitats, although impacts on fish communities can be difficult to assess.

In streams where existing infrastructures cannot be removed or modified, it is also possible to optimize sediment and water management in order to maximize connectivity and achieve flow patterns that ensure minimum ecosystem requirements. This can include releases from dams, but also delaying and/or treating water from agricultural and urban sources.

Implementing a Minimum Stream Corridor Width

Another method of ensuring the ecological health of streams while limiting impacts on human infrastructures is to delineate a corridor within which the stream is expected to migrate over time. This method is based on the concept of minimum intervention within this corridor, whose limits should be determined based on the stream's hydrology and geomorphology. Although this concept is often restricted to the lateral mobility of streams (related to bank erosion), some systems also integrate the space necessary for floods of various return periods. This concept has been developed and adapted in various countries around the world, resulting in the notion of "stream corridor" or "river corridor" in the USA, "room for the river" in the Netherlands, "espace de liberté" ("freedom space") in France (where the concept of "erodible corridor" is also used") and Québec (Canada), "espace réservé aux eaux" ("space reserved for water(courses)") in Switzerland, "fascia di pertinenza fluviale" in Italy, "fluvial territory" in Spain and "making space for water" in the United Kingdom. A cost-benefit analysis has shown that this approach could be beneficial in the long term due to lower stream stabilization and maintenance costs, lower damages resulting from erosion and flooding, and ecological services redered by the restored streams. However, this approach cannot be implemented alone if watershed-scale stressors contribute to stream degradation.

Additional Practices

In addition to the aforementioned restoration approaches and methods, additional measures can be implemented if stream degradation factors occur at the watershed scale. First, high quality areas should also be protected. Additional measures include revegetation/reforestation efforts (ideally with native species); the adoption of agricultural best management practices that minimize erosion and runoff; adequate treatment of sewage water and industrial discharge across the watershed; and improved stormwater management to delay/minimize the transport of water to the stream and minimize pollutant migration. Alternative stormwater management facilities include the following options:

- Bioretention systems and rain gardens.

- Constructed wetlands.
- Infiltration basins.
- Retention basins.

A rain garden in Singapore.

Wildlife Corridor

A green forest corridor in Brazil.

A wildlife corridor, habitat corridor, or green corridor is an area of habitat connecting wildlife populations separated by human activities or structures (such as roads, development, or logging). This allows an exchange of individuals between populations, which may help prevent the negative effects of inbreeding and reduced genetic diversity (via genetic drift) that often occur within isolated populations. Corridors may also help facilitate the re-establishment of populations that have been reduced or eliminated due to random events (such as fires or disease).

This may potentially moderate some of the worst effects of habitat fragmentation, wherein urbanization can split up habitat areas, causing animals to lose both their natural habitat and the ability to move between regions to use all of the resources they need to survive. Habitat fragmentation due to human development is an ever-increasing threat to biodiversity, and habitat corridors are a possible mitigation.

Purpose

The main goal of implementing habitat corridors is to increase biodiversity. When areas of land are broken up by human interference, population numbers become unstable and many animal and plant species become endangered. By re-connecting the fragments, the population fluctuations can decrease dramatically. Corridors can contribute to three factors that stabilize a population:

- Colonization: Animals are able to move and occupy new areas when food sources or other natural resources are lacking in their core habitat.
- Migration: Species that relocate seasonally can do so more safely and effectively when it does not interfere with human development barriers.
- Interbreeding: Animals can find new mates in neighboring regions so that genetic diversity can increase and thus have a positive impact on the overall population.

The definitions of "biological corridor" (i.e., wildlife corridor) had, in the early years of studying corridors, been "vague and inconsistent, and often they confound form and function" Rosenberg developed a conceptual model that emphasized the role of a wildlife corridor as a facilitator of movement that is not restricted by requirements of native vegetation or intermediate target patches of habitat. Their definition simply required that movement to a target patch via the corridor be greater that if the corridor were absent.

Although corridors had originally been implemented with the assumption that they would increase biodiversity, not enough research had been done to come to a solid conclusion. The case for corridors has been built more on intuition and much less on empirical evidence. The early controversies had arisen because most studies had been limited in that they had a narrow taxonomic focus and, that if corridors facilitate animal movement, they should also have strong indirect effects on plant populations due to increased pollen and seed by animals. Results of their 2002 experiment provided a large-scale experimental demonstration that habitat (or wildlife) corridors facilitate movement of disparate taxa between otherwise isolated patches even after controlling for area effects. Another factor that needs to be taken into account is what species the corridor is intended for. Some species have reacted more positively to corridors than others.

A habitat corridor could be considered as a possible solution in an area where the destruction of a natural area has greatly affected its native species. Development such as roads, buildings, and farms can interrupt plants and animals in the region being destroyed. Furthermore, natural disasters such as wildfires and floods can leave animals with no choice but to evacuate. If the habitat is not connected to a safer one, it will ultimately lead to death. A remaining portion of natural habitat is called a remnant, and such portions need to be connected because when migration decreases, extinction increases.

Corridors can be made in two distinct areas either water or land. Water corridors are called riparian ribbons and usually come in the form of rivers and streams. Land corridors come on a scale as large as wooded strips connecting larger woodland areas. However, they can also be as simple as a line of shrubs along a sidewalk. Such areas can facilitate the movement of small animals, especially birds, from tree to tree, until they find a safe habitat to rest in. Not only do minimal corridors aid in the movement of animals, they are also aesthetically pleasing, which can sometimes encourage the community to accept and support them.

Users

Species can be categorized in one of two groups; passage users and corridor dwellers.

Passage users occupy corridors for brief periods of time. These animals use corridors for such events as seasonal migration, dispersal of a juvenile, or moving between parts of a large home range. Usually large herbivores, medium to large carnivores, and migratory species are passage users. One common misconception is that the corridor only needs to be wide enough for the passage users to get through. However, the corridor still must be wide enough to be safe and also encourage the animals to use it, even though they do not live out their entire lives in it.

Corridor dwellers can occupy the passage anywhere from several days to several years. Species such as plants, reptiles, amphibians, birds, insects, and small mammals can spend their entire lives in linear habitats. In this case, the corridor must include everything that a species needs to live and breed, such as soil for germination, burrowing areas, and multiple other breeding adults.

Types

Habitat corridors can be categorized according to their width. Typically the wider the corridor, the more use it will get from species. However, the width-length ratio, as well as design and quality play just as important of a role in creating the perfect corridor. The strip of land will suffer less from edge effects such as weeds, predators, and chemicals if it is constructed properly. The following are three divisions in corridor widths:

- Regional: (>500m wide); connect major ecological gradients such as migratory pathways.
- Sub-regional: (>300m wide); connect larger vegetated landscape features such as ridgelines and valley floors.
- Local: (some <50m); connect remnant patches of gullies, wetlands, ridgelines, etc.

Habitat corridors can also be divided according to their continuity. Continuous corridors are strips that are not broken up, while "stepping stone" corridors are small patches of suitable habitat. When stepping stones are arranged in a line, they form a strip of land connecting two areas, just like a continuous corridor would. Both kinds provide linkages between protected core areas and stimulate or allow species to migrate.

Wildlife overpass in Singapore.

Wildlife underpass for bears in Florida, United States.

Finally, corridors can come in the form of underpasses or overpasses, which can be very safe for both animals and humans. Many busy highways cross through natural habitats that native species occupy, as well. Large animals such as deer become a hazard when they cross in front of traffic and get hit. An overpass or an underpass serves as a bridge to facilitate the movement of animals across a busy road. Observations have shown that underpasses are actually more successful than overpasses because many times animals are too timid to cross over a bridge in front of traffic and would prefer to be more hidden.

Costs

Corridors can be expensive to plan out and put into action. For example, "a bridge that would maintain a riparian corridor costs about 13 times as much per lane-mile as would a road that would sever the corridor." He also states that maintenance of a corridor would be much more costly than refuges for endangered species. It would simply be easier to move animals between refuges than to buy land, install a corridor, and maintain it. However, where the goal is not just to preserve a few large animal species but to protect biodiversity among all plants and animals, then habitat corridors may be the only option. Corridors are going to be expensive to implement no matter what, but it does depend on the type, location, and size, which can all vary to a great degree. With the lack of field data on the effectiveness, many agencies are not willing to consider putting in corridors.

Monitoring Use

It is extremely important for researchers to pay attention to the population changes in animals after a corridor has been implemented to ensure that there are no harmful effects. Researchers can use both mark-recapture techniques and evaluate genetic flow in order to observe how much a corridor is being used. Marking and recapturing animals is more useful when keeping a close eye on individual movement. The only problem is that tagging animals and watching them does not tell anyone whether the migrating individuals are successfully mating with other populations in connected areas of land. On the other hand, genetic techniques can be more effective in evaluating migration and mating patterns.

One of the most important goals of developing a corridor is to increase migration in certain animal species. By looking at a population's gene flow, researchers can understand the genetic consequences of corridors. The migration patterns of an entire population are much more important than the movements of a few individuals. From these techniques, researchers will better understand whether or not habitat corridors are increasing biodiversity.

Stephen Mech and James Hallett introduce an additional reason genetic techniques are more useful; they "measure average migration rates over time, which reveals the effects of fragmentation of several generations and is not as sensitive to current population sizes as mark-recapture studies are." For example, when a population is extremely small, mark-recapture is almost impossible. Clearly, genetic analysis of a species is the best way to determine if animals are actually using corridors to move and reproduce.

Design

According to new research, wildlife corridors are best built with a certain degree of randomness or asymmetry, rather than built symmetrically. The research was conducted at UC Davis.

Wildlife corridors are susceptible to edge effects; habitat quality along the edge of a habitat fragment is often much lower than in core habitat areas. Wildlife corridors are important for large species requiring significant sized ranges; however, they are also vital as connection corridors for smaller animals and plants as well as ecological connectors to provide a rescue effect.

Both the safety of animals and humans can be achieved through the creation of corridors. For example, deer commonly cross roads in order to get to other grazing land. When they are faced with a car coming at them, they freeze; this puts both the deer and the human's life in danger. In Alberta, Canada, an overpass was constructed to keep animals off of the busy highway; the area is part of a national park, so many different creatures roam the area. The top of the bridge is covered in the native grass of the area so that it blends in better and animals will not know the difference. Gates were also put of on either side of the overpass to help guide animals in the right direction.

In Southern California, 15 underpasses and drainage culverts were observed to see how many animals used them as corridors. They proved to be especially effective on wide-ranging species such as carnivores, mule deer, small mammals, and reptiles, even though the corridors were not intended specifically for animals. Researchers also learned that factors such as surrounding habitat, underpass dimensions, and human activity also played a role in how much use they got. From this experiment, much was learned about what would constitute a successful habitat corridor.

In South Carolina, five remnant areas of land were monitored; one was put in the center and four were surrounding it. Then, a corridor was put between one of the remnants and the center. Butterflies that were placed in the center habitat were two to four times more likely to move to the connected remnant rather than the disconnected ones. Furthermore, male holly plants were placed in the center region, and female holly plants in the connected region increased by 70 percent in seed production compared to those plants in the disconnected region. The most impressive dispersal into the connected region, though, was through bird droppings. Far more plant seeds were dispersed through bird droppings in the corridor-connected patch of land.

There have also been positive effects on the rates of transfer and interbreeding in vole populations. A control population in which voles were confined to their core habitat with no corridor was compared to a treatment population in their core habitat with passages that they could use to move to other regions. Females typically stayed and mated within their founder population, but the rate of transfer through corridors in the males was very high. Researchers are not sure why the females did not move about as much, but it is apparent that the corridor effectively transferred at least some of the species to another location for breeding.

In 2001, a wolf corridor was restored through a golf course in Jasper National Park, Alberta, which enabled wolves to pass through the course. After this restoration, wolves passed through the corridor frequently. This is one of the first demonstrations that corridors are used by wildlife, and can be effective in decreasing fragmentation. Earlier studies had been criticised for failing to demonstrate that corridor restoration leads to a change in wildlife behaviour.

Elephant Corridor

Elephant corridors are narrow strips of land that allow elephants to move from one habitat patch to another. There are 88 identified elephant corridors in India.

In Africa, Botswana houses the largest number of free-roaming elephant herds. Elephants Without Borders (EWB) studies the movement of elephants is working to gain community support of local community corridors, so that elephants and humans can co-exist.

Major Wildlife Corridors

Several artificial wildlife corridors have been planned or created, these include:

- The Paséo Pantera (also known as the MesoAmerican Biological corridor or Paséo del Jaguar).
- The Eastern Himalayan Corridor.
- China-Russia Tiger Corridor.
- Tandai Tiger Corridor.
- The European Green Belt.
- The Siju-Rewak Corridor, located in the Garo Hills of India, protects an important population of elephants(thought to be approximately 20% of all the elephants that survive in the country).This corridor project links together the Siju Wildlife Sanctuary and the Rewak Reserve Forest in Meghalaya State, close to the India-Bangladesh border. This area lies within the meeting place of the Himalayan Mountain Range and the Indian Peninsula and contains at least 139 other species of mammal, including tiger, clouded leopard and the Himalayan black bear.

- The Ecologische Hoofdstructuur is a network of corridors and habitats created for wildlife in the Netherlands.

Evaluation

Some animal species are much more apt to use habitat corridors than others depending on what their migration and mating patterns are like. For example, many cases of birds and butterflies successfully using corridors have been observed. Less successful stories have come out of mammals such as deer. How effective a corridor is may simply rely on what species it is directed towards. Corridors created with birds in mind may be more successful because they are highly migratory to begin with.

Human interference is almost inevitable with the quickly increasing population. The goal behind habitat corridors shows the most hope for solving habitat fragmentation and restoring biodiversity as much as possible. Although there are many positives and negatives, there may be enough positives to continue studying and improving corridors. It is truly difficult to say whether corridors are the solution to increasing biodiversity, because each one must be judged on its own. Each corridor has its own set of standards and goals that may set it apart from another one.

Negatives

A major downfall to habitat corridors is that not much information has been gathered about their success. Due to the lack of positive data, many agencies will not allow corridors to be established because they are unsure of their effectiveness. Another problem with corridors is that they are not as useful as simply preserving land so that it cannot be fragmented. However, it is becoming very difficult to set aside land for nature reserves when road-building, industry, and urban sprawl are all competing for space.

Even if corridors are sought as a solution, it does not necessarily mean that animals will use them. Especially in the case of overpasses, research shows that animals do not like to use them to get to another remnant area of land. Usually overpasses are built over busy highways, and many species are too timid to expose themselves in front of all of the traffic. As more roads and buildings arise, there becomes less space to try to preserve.

Habitat corridors need to be species-specific (not every kind of animal will use every kind of corridor) and corridors can be barriers to some species. For instance plants may use road verges as corridors however some mammals will not cross roads to reach a suitable habitat.

When a corridor is implemented, many times development is so close by, that it becomes difficult to build a wide enough passage. There is usually a very limited amount of space available for corridors, so buffers are not usually added in. Without a buffer zone, corridors become susceptible to harmful outside factors from city streets, suburb development, rural homes, forestry, cropland, and feedlots.

Unfortunately, another limiting factor to the implementation of corridors is money. With such inconclusive data about the effectiveness of connecting land, it is difficult to get the proper funding. Those who would be in charge of the corridor design and construction would ask such questions as, "What if the corridors affect species negatively?" and "What if they actually aid in the spread of disease and catastrophic events?" Furthermore, there is a possibility that corridors could not only aid in the dispersal of native organisms, but invasive ones, as well. If invasive species take over an area they could potentially threaten another species, even to the point of extinction.

Although wildlife corridors have been proposed as solutions to habitat and wildlife population fragmentation, there is little evidence that they are broadly useful as a conservation strategy for all biodiversity in non-developed or less-developed areas, compared to protecting connectivity as the relevant ecological attribute. In other words, corridors may be a useful meme for conservation planning/ers, but the concept has less meaning to wildlife species themselves. Very few wildlife follow easily identified "corridors" or "linkages" (e.g., using computer modeling), instead most species meander and opportunistically move through landscapes during daily, seasonal, and dispersal movement behavior. Wildlife corridors may be useful in highly developed landscapes where they are easily identified as the last remaining and available habitat.

Positives

Habitat corridors may be defenseless against a number of outside influences, but they are still an efficient way of increasing biodiversity. Strips of land aid in the movement of various animal species and pollen and seed dispersal, which is an added benefit to the intended one. For example, when insects carrying pollen or birds carrying seeds travel to another area, plant species effectively get transported, as well.

Another positive aspect of corridors is that they allow both animals and humans to occupy virtually the same areas of land, and thus co-exist where without the corridor this would not be possible. Large animals such as bears can be attracted to residential areas in search of food due to lack of natural resources because of habitat fragmentation. A corridor would provide a passage for the bears to forage in other locations, so that they would not pose as much of a threat to humans.

Prairie Restoration

Prairie restoration is a conservation effort to restore prairie lands that were destroyed due to industrial, agricultural, commercial, or residential development. For example, the U.S. state of Illinois alone once held over 35,000 square miles (91,000 km^2) of prairie land and now just 3 square miles (7.8 km^2) of that original prairie land is left.

A tallgrass prairie planting in Illinois.

Purpose

Ecologically, prairie restoration aids in conservation of earth's topsoil, which is often exposed to erosion from wind and rain when prairies are plowed under to make way for new commerce. Conversely, much more of the prairie lands have become the fertile fields on which cereal crops of corn, barley and wheat are grown.

Many prairie plants are also highly resistant to drought, temperature extremes, disease, and native insect pests. They are frequently used for xeriscaping projects in arid regions of the American West.

A Coastal prairie restoration project in Texas.

A restoration project of prairie lands can be large or small. A backyard prairie restoration will enrich soil, help with erosion and take up extra water in excessive rainfalls. Prairie flowers are attractive to native butterflies and other pollinators. On a larger scale, communities and corporations are creating areas of restored prairies which in turn will store organic carbon in the soil and help maintain the biodiversity of the 3000 plus species that count on the grasslands for food and shelter.

Types of Plants

Some prominent tallgrass prairie grasses include big bluestem, indiangrass, and switchgrass. Midgrass and shortgrass species include little bluestem, side oats grama, and buffalograss. Many of the diverse prairie forbs (herbaceous, non-graminoid flowering plants) are structurally specialized to resist herbaceous grazers such as American bison. Some have hairy leaves that may help deter the cold and prevent excessive evaporation. Many of forbs contain secondary compounds that were discovered by indigenous peoples and are still used widely today. For example, purple coneflower (Echinacea purpurea or just Echinacea) is used as an herbal remedy for colds.

Early prairie restoration efforts tended to focus largely on a few dominant species, typically grasses, with little attention to seed source. With experience, later restorers have realized the importance of obtaining a broad mix of species and using local ecotype seed.

Care of Prairies

Fire is a big component to the success of grasslands, large or small. Controlled burns, with a permit, are recommended every 4–8 years (after two growth seasons) to burn away dead plants; prevent certain other plants from encroaching (such as trees) and release nutrients into the ground to encourage new growth. A much more wildlife habitat friendly alternative to burning every 4–8 years is to burn 1/4 to 1/8 of a tract every year. This will leave wildlife a home every year and still accomplish the task of burning. The Native Americans may also have used the burns to control pests such as ticks. If controlled burns are not possible, rotational mowing is recommended as a substitute.

One of the newer methods available is holistic management, which uses livestock as a substitute for the keystone species such as bison. This allows the rotational mowing to be done by animals which in turn mimics nature more closely. Holistic management also can use fire as a tool, but in a more limited way and in combination with the mowing done by animals.

Mangrove Restoration

Mangrove restoration is the regeneration of mangrove forest ecosystems in areas where they have previously existed. The practice of mangrove restoration is grounded in the discipline of restoration ecology, which aims to "[assist] the recovery of resilience and adaptive capacity of ecosystems that have been degraded, damaged, or destroyed". Since environmental impacts are an ongoing threat, to successfully restore an ecosystem implies not merely to recreate its former condition, but to strengthen its capacity to adapt to change over time.

Mangrove forests, along with the animal species they shelter, represent globally significant sources of biodiversity and provide humanity with valuable ecosystem services. They are used by mammals, reptiles and migratory birds as feeding and breeding grounds, and provide crucial habitats for fish and crustacean species of commercial importance. The roots of the mangrove physically buffer shorelines from the erosive impacts of ocean waves and storms. Additionally, they protect riparian zones by absorbing floodwaters and slowing down the flow of sediment-loaded river water. This allows sediments to drop to the bottom where they are held in place, thus containing potentially toxic waste products and improving the quality of water and sanitation in coastal communities.

To the human communities who rely on them, mangrove forests represent local sources of sustainable income from the harvest of fish and timber, as well as non-timber forest products such as medicinal plants, palm leaves and honey. On a global scale, they have been shown to sequester carbon in quantities comparable to higher-canopy terrestrial rainforests, which means that they may play a role in climate change mitigation, in addition to physically protecting coastlines from the projected sea-level rise associated with climate change. However, there are limits to the capacity of mangroves to adapt to climate change. It is projected that a 1-meter rise in sea level could inundate and destroy mangrove forests in many regions around the globe, which would leave coastal communities vulnerable to the risks of flooding, shoreline erosion, saline intrusion and increased storm activity.

The issue of restoration is critical today since mangrove forests are being lost very quickly – at an even faster rate than tropical rainforests inland. A recent estimate puts the total mangrove area worldwide in 2005 at 152,000 km^2 down from 188,000 km^2 in 1980. In other words, some 36,000 km^2, or nearly 20% of the world's mangroves, were lost over a period of twenty-five years. Other estimates of loss may differ due to having been drawn from a smaller pool of data. The Millennium Ecosystem Assessment estimates the total loss worldwide at 35% between 1980 and 2000, but this result was drawn from data on only slightly more than half of the total mangrove area. Much of this lost mangrove area was destroyed to make room for industry, housing and tourism development; for aquaculture, primarily shrimp farms; and for agriculture, such as rice paddies, livestock pasture and salt production. Other drivers of mangrove forest destruction include activities that divert their sources of freshwater, such as groundwater withdrawals, the building of dams, and the building of roads and drainage canals across tidal flats.

Mangroves are sensitive ecosystems, changing dynamically in response to storms, sediment blockage, and fluctuations in sea level and present a “moving target” for restoration efforts. Different restoration approaches face this challenge in different ways. The most common method simply consists in planting single-species stands of mangroves in areas thought to be suitable, without consideration of whether or not they supported mangroves in the past. This approach usually fails over the long term because the underlying soil and hydrological requirements of the mangroves are not being met. More informed methods aim to bring a damaged mangrove area back into its preexisting condition, taking into account not only ecosystem factors but also social, cultural and political perspectives. These approaches begin with the understanding that a damaged mangrove area may be able to repair itself through the natural processes of secondary succession, without being physically planted, provided that its tidal and freshwater hydrology is functioning normally and there is an adequate supply of seedlings. Taking this into account, it becomes crucial to the success of a restoration project to evaluate what the hydrology of a disturbed mangrove site should look like under normal conditions, and the ways in which it has been modified. One example of this approach is the Ecological Mangrove Restoration method which recommends the following steps, to be undertaken using healthy mangroves of the surrounding area as a reference:

- Assess the ecology, especially reproduction and distribution patterns, of the mangrove species at the disturbed site.
- Map the topographical elevations and hydrological patterns that determine how seedlings should establish themselves at the site.
- Assess the changes made to the site that currently prevent the site from recovering by itself.
- Design a restoration plan that begins by restoring the normal range of elevations and tidal hydrology at the site.
- Monitor the site to determine if the restoration has been successful in light of the original objectives.

The actual planting of seedlings is a last resort, since it fails in many cases; it should be considered only if natural recruitment of seedlings fails to reach the restoration objective.

Mangroves as Climate Change Mitigation

Mangrove forests have a potential to mitigate climate change, such as through the sequestration of carbon from the atmosphere directly, and by providing protection from storms, which are expected to become more intense and frequent into the 21st century. Wetland plants, like mangroves, take in carbon dioxide when they perform photosynthesis. They then convert this into biomass made of complex carbon compounds. Being the most carbon-rich tropical forest, mangroves are highly productive and are found to store 3 to 4 times more carbon than other tropical forests. This is known as Blue carbon. Mangroves make up only 0.7% of tropical forest area worldwide, yet studies calculate the effect of mangrove deforestation to contribute 10% of global CO_2 emissions from deforestation. The image to the right shows the global distribution of above ground carbon from mangroves. As can be seen, most of this carbon is located in Indonesia, followed by Brazil, Malaysia and Nigeria. Indonesia has one of the highest rates of mangrove loss, yet the most carbon stored from mangroves. Therefore, it is suggested that if the correct policy is implemented, countries like Indonesia can make considerable contributions to global carbon fluxes.

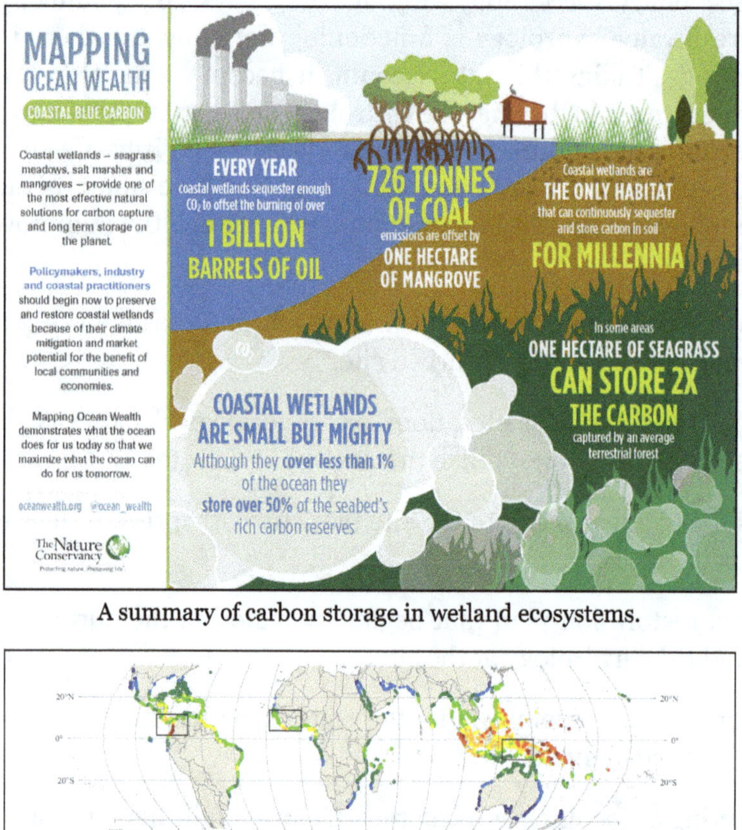

A summary of carbon storage in wetland ecosystems.

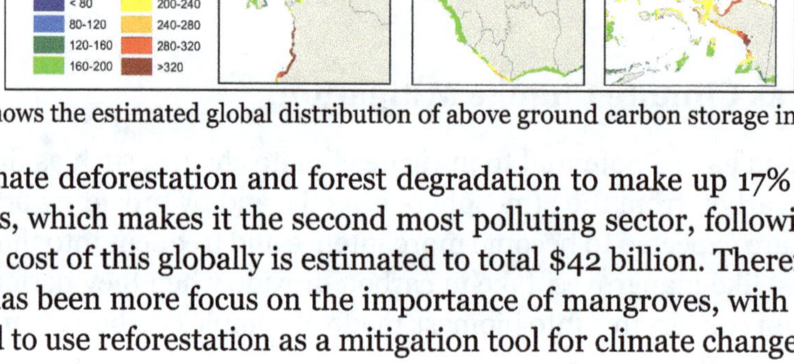

This map shows the estimated global distribution of above ground carbon storage in mangroves.

The UN estimate deforestation and forest degradation to make up 17% of global carbon emissions, which makes it the second most polluting sector, following the energy industry. The cost of this globally is estimated to total $42 billion. Therefore, in recent years, there has been more focus on the importance of mangroves, with initiatives being developed to use reforestation as a mitigation tool for climate change.

Reducing Emissions from Deforestation and Forest Degradation

In 2008, the United Nations launched the "Reducing Emissions from Deforestation and forest Degradation (REDD)" program to combat climate change through the reduction of carbon emissions and enhancement of carbon sinks from forests. It is the opinion of literary scholars that the REDD program can increase carbon sequestration from mangroves and therefore reduce carbon in the atmosphere. The REDD+ mechanism, as part of the REDD program, provides financial support to stakeholders in developing countries to avoid deforestation and forest degradation. The estimated impacts of REDD+ globally, could reach up to 2.5 billion tons of CO_2 each year. An examples of REDD+ implementation can be seen in Thailand, where carbon markets give farmers incentive to conserve mangrove forests, by compensating for the opportunity cost of shrimp farming.

Mangroves for the Future

Moreover, the Mangroves for the Future (MFF) initiative, led by IUCN and UNDP, encourages the rehabilitation of mangroves by engaging with local stakeholders and creating a platform for change. In Indonesia, one project planted 40,000 mangroves, which then encouraged local government to take up similar initiatives on a larger scale. Mangrove restoration and protection is also seen as a climate change mitigation strategy under COP21, the international agreement to target climate change, with countries being able to submit the act in their Nationally Appropriate Mitigation Approaches (NAMAs). Ten of the world's least developed countries are now prioritizing mangrove restoration in their NAMAs.

Climate Change Adaptation

As well as providing the benefit of a carbon sinks, mangroves also host potential for climate change adaptation. They provide protection to local communities from sea level rise, coastal erosion and storms. These are all issues that are related to climate change and are expected to increase in severity in the future. Therefore, mangroves can help support the livelihood of those living in areas already vulnerable to climate change threats. In the IPCC AR5 report, the potential of ecosystem-based adaptation (EBA) to climate change is discussed, which includes the restoration of mangroves. An example of this can be seen in Bangladesh, where the government initiated the plantation of 50,000 hectares of mangrove forest to stabilize coastal areas, in an attempt to tackle increasing erosion. Evidence suggests that this initiative was successful in increasing accretion of coastal sediment, which reduced coastal erosion in this area and protects coastal communities from flooding and storm events. It has also been found that areas surrounding mangrove forests are subject to less damage from cyclones than non forested areas.

Additional Considerations

An important but often overlooked aspect of mangrove restoration efforts is the role that the local communities play as stakeholders in the process and the outcome. Since they may directly feel the effects of restoration projects, they should be involved in the process as much as feasibly possible, from decision-making to maintenance over the long term. Their involvement and local knowledge, as well as collaboration with other stakeholders such as sponsors and governing agencies, is crucial to the success of restoration projects.

Limitations

In some areas, restoration may be prohibitively difficult due to the degradation of the soil that regularly follows the clear-cutting of mangrove forests. Common effects include advanced erosion of the soil, loss of nutrients, high levels of salinity, and/or buildup of toxins. However, even without this extent of degradation, the soil may become unable to host plant life at all due to the loss of the live mangrove roots, which exuded oxygen and carbohydrate into the soil and maintained its quality. Using foresight early in the restoration process to carefully select sites that are likely to succeed as self-maintaining ecosystems, as well as ensuring that proper management is built into the conservation effort, can prevent the waste of time and energy that often accompanies restoration projects. The long-lasting aftereffects of mangrove degradation underscore the importance of eliminating its causes, since once sites are cleared, it is difficult for them to recover without a scientific intervention.

Soil Seed Bank

Soil seed bank refers to the natural storage of seeds in the leaf litter, on the soil surface, or in the soil of many ecosystems, which serves as a repository for the production of subsequent generations of plants to enable their survival. The term soil seed bank can be used to describe the storage of seeds from a single species or from all the species in a particular area. Given the variety of stresses that ecosystems experience such as cold, wildfire, drought, and disturbance seed banks are often a crucial survival mechanism for many plants and maintain the long-term stability of ecosystems.

The Role of Seed Dormancy

Seed dormancy and environmental constraints on germination influence various characteristics of soil seed banks. For example, seed dormancy determines how long a seed can remain viable in the soil. Factors such as embryo immaturity, chemical inhibitors, and physical constraints influence seed dormancy. Light filtered through plant canopies, for example, can inhibit germination in some species, while a long winter chilling may break dormancy in other species. The result is a considerable variety in the patterns of germination of the seed banks by seasons, disturbances, or other environmental shifts.

Variation in the characteristics of seed dormancy determine whether a species's soil seed bank is transient (temporary) or persistent. Transient seed banks are composed of species that produce seeds with a brief or no period of dormancy. Such seeds generally germinate prior to the next round of seed production, and the seed bank is thus continually depleted and reestablished. Transient seed banks are typical for many plants, especially long-lived perennials such as trees and shrubs. Often, such species rely on other strategies or life-history stages for persistence. For example, species may depend on long-lived adults, "banks" of seedlings in a forest understory, or extensive seed dispersal. In contrast, species with persistent seed banks have seeds that can remain dormant for more than a year, meaning that there is always some viable seed in the soil as a reserve. Persistent seed banks are common in annual plants and some woody plants, in which the failure of seed to establish the next generation would mean the collapse of the population. Scientists sometimes further classify persistent seed banks based on the extent or pattern of dormancy.

The Role of Disturbance

In addition to dormancy, considerable variation occurs in seed bank germination because of seasonal or other environmental shifts. Disturbances such as fire, flooding, windstorms, plowing, or forest clearing are frequently strong selective forces and may increase the overall germination response of seeds. Ecosystems characterized by wildfire often have extreme cases of persistent seed banks, as is common for many areas with Mediterranean climates, such as Australia, California, and South Africa. In those ecosystems the germination of many species requires signals provided by fire, such as a heat pulse into the soil or chemicals from smoke or charred wood. Germination may not occur until after a wildfire, which then results in mass germination from the seed bank the following spring. Similarly, the seed banks of agricultural weeds are often well adapted to the almost continuous human-made disturbances of their environment. Such weeds frequently have complex dormancy patterns that reflect the agricultural practices under which they evolved.

Seed Bank Modeling

Researcher Dan Cohen was one of the first scientists to model soil seed banks. In the 1960s, focusing on desert annuals subject to highly irregular rainfall, he developed population-dynamics models that suggested that a reserve of some fraction of seed in the soil was essential for the plants to avoid local extinction. Cohen found that the dynamics of soil seed banks reflect the degree of ecological constraint a species or population faces in establishing the next generation. Although his work focused on annuals, the conceptual framework applies readily to any plant species. Such modeling is important to ecological research and conservation planning, as traditional demographic models and field surveys often fail to consider population reserves in the soil.

References

- Restoration-ecology-13339059: nature.com, Retrieved 16 June, 2020
- Reichenauer TG, Germida JJ (2008). "Phytoremediation of organic contaminants in soil and groundwater". ChemSusChem. 1 (8–9): 708–17. doi:10.1002/cssc.200800125. PMID 18698569
- Soil-seed-bank, science: britannica.com, Retrieved 14 February, 2020
- Stenhouse (2004). "Local government conservation and management of native bushland in Australia". Environmental Management. Springer Scientific. 34 (2): 209–222. doi:10.1007/s00267-004-0231-6
- Chao, A.; Tsay, P. K.; Lin, S. H.; Shau, W. Y.; Chao, D. Y. (2001). "The applications of capture-recapture models to epidemiological data". Statistics in Medicine. 20 (20): 3123–3157. doi:10.1002/sim.996. PMID 11590637
- Tewksbury, Joshua (1 October 2002). "Corridors affect plants, animals, and their interactions in fragmented landscapes". Proceedings of the National Academy of Sciences of the United States of America. 99 (20): 12923–6. Bibcode:2002PNAS...9912923T. doi:10.1073/pnas.202242699. PMC 130561. PMID 12239344. Retrieved 14 September, 2018

Permissions

All chapters in this book are published with permission under the Creative Commons Attribution Share Alike License or equivalent. Every chapter published in this book has been scrutinized by our experts. Their significance has been extensively debated. The topics covered herein carry significant information for a comprehensive understanding. They may even be implemented as practical applications or may be referred to as a beginning point for further studies.

We would like to thank the editorial team for lending their expertise to make the book truly unique. They have played a crucial role in the development of this book. Without their invaluable contributions this book wouldn't have been possible. They have made vital efforts to compile up to date information on the varied aspects of this subject to make this book a valuable addition to the collection of many professionals and students.

This book was conceptualized with the vision of imparting up-to-date and integrated information in this field. To ensure the same, a matchless editorial board was set up. Every individual on the board went through rigorous rounds of assessment to prove their worth. After which they invested a large part of their time researching and compiling the most relevant data for our readers.

The editorial board has been involved in producing this book since its inception. They have spent rigorous hours researching and exploring the diverse topics which have resulted in the successful publishing of this book. They have passed on their knowledge of decades through this book. To expedite this challenging task, the publisher supported the team at every step. A small team of assistant editors was also appointed to further simplify the editing procedure and attain best results for the readers.

Apart from the editorial board, the designing team has also invested a significant amount of their time in understanding the subject and creating the most relevant covers. They scrutinized every image to scout for the most suitable representation of the subject and create an appropriate cover for the book.

The publishing team has been an ardent support to the editorial, designing and production team. Their endless efforts to recruit the best for this project, has resulted in the accomplishment of this book. They are a veteran in the field of academics and their pool of knowledge is as vast as their experience in printing. Their expertise and guidance has proved useful at every step. Their uncompromising quality standards have made this book an exceptional effort. Their encouragement from time to time has been an inspiration for everyone.

The publisher and the editorial board hope that this book will prove to be a valuable piece of knowledge for students, practitioners and scholars across the globe.

Index

www.ingramcontent.com/pod-product-compliance
Lightning Source LLC
LaVergne TN
LVHW080612210926
848485LV00003B/143

* 9 7 8 1 6 3 9 8 9 4 3 6 9 *